SEISMIC DESIGN FOR ARCHITECTS
OUTWITTING THE QUAKE

SEISMIC DESIGN FOR ARCHITECTS

OUTWITTING THE QUAKE

Andrew Charleson

Routledge
Taylor & Francis Group

LONDON AND NEW YORK

Architectural Press is an imprint of Routledge

First edition 2008

2 Park Square, Milton Park, Abingdon, Oxon OX14 4RN
711 Third Avenue, New York, NY 10017, USA

Routledge is an imprint of the Taylor & Francis Group, an Informa business

Notice
No responsibility is assumed by the publisher for any injury and/or damage to persons
or property as a matter of products liability, negligence or otherwise, or from any use
or operation of any methods, products, instructions or ideas contained in the material
herein.

British Library Cataloguing in Publication Data
A catalogue record for this book is available from the British Library

Library of Congress Cataloguing in Publication Data
A catalogue record for this book is available from the Library of Congress

ISBN: 978-0-7506-8550-4

For information on all Architectural Press publications
visit our website at www.routledge.com

Contents

Foreword by Christopher Arnold, FAIA, RIBA   ix

Preface   xi

Acknowledgements   xiii

1   Earthquakes and ground shaking   1

Introduction   1
Understanding earthquakes   4
Earthquake magnitude and intensity   9
The nature of earthquake shaking   11
Importance of ground conditions   13
References and notes   14

2   How buildings resist earthquakes   15

Introduction   15
Nature of seismic forces   15
Factors affecting the severity of seismic forces   18
Resisting seismic forces   25
Torsion   27
Force paths   29
Notes   32

3   Seismic design approaches   33

Introduction   33
Historical overview   33
Current seismic design philosophy   38
References and notes   47

4   Horizontal structure   49

Introduction   49
Diaphragms   50
Transfer diaphragms   56
Bond beams   58
Collectors and ties   61
Note   61

5 Vertical structure 63

Introduction 63
Shear walls 66
Braced frames 76
Moment frames 81
Mixed systems 89
References 91

6 Seismic design and architecture 93

Introduction 93
Integrating seismic resisting structure and architecture 94
How much structure is needed? 99
Special structures 102
Contemporary architecture in seismic regions 104
Case study: the Villa Savoye 108
References and notes 112

7 Foundations 113

Introduction 113
Seismic foundation problems and solutions 114
Foundation types 119
Foundation investigations 119
Retaining structures 121
References and notes 123

8 Horizontal configuration 125

Introduction 125
Torsion 128
Re-entrant corners 132
Diaphragm discontinuities 134
Non-parallel systems 136
Pounding and separation 137
Bridging between buildings 140
References and notes 141

9 Vertical configuration 143

Introduction 143
Soft storeys 144
Short columns 148
Discontinuous and off-set walls 151
Setbacks 154

Buildings on sloping sites 155
References and notes 155

10 Non-structural elements: those likely to cause structural damage 157

Introduction 157
Infill walls 159
Staircases 168
References 171

11 Other non-structural elements 173

Introduction 173
Cladding 174
Parapets and appendages 181
Partition walls 182
Suspended ceilings and raised floors 182
Mechanical and electrical equipment 184
Building contents 184
References 186

12 Retrofitting 187

Introduction 187
Why retrofit? 189
Retrofit objectives 191
Retrofit approaches 192
Retrofit techniques 195
Non-structural retrofit 202
Historic buildings 203
References 204

13 Professional collaboration and communication 207

Introduction 207
Client 208
Design team 210
Contractor 213
Post-earthquake 215
References and notes 216

14 New technologies 217

Introduction 217
Seismic isolation 218

Dampers 224
Damage avoidance 227
Innovative structural configurations 228
Structural design approaches 229
Other developments 230
References 231

15 Urban planning **233**

Introduction 233
Planning 234
Tsunami 237
Fire following earthquake 238
Interdisciplinary interaction 240
References and notes 240

16 Issues in developing countries **243**

Introduction 243
Design 245
Construction 248
Resources 248
References 249

17 Earthquake architecture **251**

Introduction 251
Expression of seismic resistance 253
Expression of structural principles and actions 255
Seismic issues generating architecture 258
References and notes 262

18 Summary **265**

Resources **269**

Introduction 269
Institutions and organizations 269
Publications 272

Index **275**

FOREWORD

I knew that I would enjoy this book when I saw that Andrew Charleson had used one of my favorite buildings, the Villa Savoie in Paris, as a seismic design case study. The earthquake engineers' nightmare, with its pin-like pilotis, ramps and roof garden – the epitome of the free planned International Style dwelling – it floats above the field in Poissy, giving the illusion of being on the sea. The author uses his re-design to demonstrate that, to add seismic resistance as an afterthought to a completed preliminary design, results in a far from elegant solution given the incompatibility of the seismic-resisting structure with the intended interior planning.

This little study is but one example of how he has made material, with which I am reasonably familiar, seem fresh and intriguing. I also liked his analogy between finger snapping and the sudden release of energy that initiates an earthquake.

Another pleasure was that in two hundred and sixty-odd pages he covers every seismic design issue under the sun with only a passing mention of seismic codes and only one (I believe) equation F=MA. The seismic codes say nothing about seismic design, which is the act of conceiving a strategy for the reduction of seismic risk and the structural/architectural systems that will accomplish it. Reading, or reading about, building codes and regulations is only one form of slow torture.

The author's intent (with which I agree) is ambitious. Structure, he says, is an *indispensable architectural element imbued with the possibility of enhancing architectural functions and qualities*, and if structure is to play architectural roles other than load-bearing, its design cannot be left to just anybody. An architect, he says, should have the skills to conceive the structural configuration at the preliminary design stage, which not only satisfies programmatic requirements and his or her design ideas, but is structurally sound with respect to seismic forces. This book is intended to provide the means by which the architect (with considerable diligence) can acquire these skills.

Such talk may, of course, upset our engineering friends (although note that the author is an engineer) and cause grumbling about the engineering ignorance of architects together with their unreasonable egotisms.

But the author is talking about **preliminary** design, the most important phase of the design process, in which all the overall configuration, the interior spaces, exterior skin, general dimensions and materials are defined. How can this be done properly without, at the same time, defining the structure? In fact, the author recommends collaboration between the engineer and architect at the earliest point in the design process. This will be more effective if the architect has a good knowledge of the structural issues.

Faced with this self-imposed task, Andrew Charleson has, I think, written a landmark book in the exposition of complex structural and architectural concept issues that use lucid prose to describe concepts and hundreds of diagrams and photographs to illuminate his message. It is instructive to discover how many sophisticated structural concepts can be explained in word and illustration to help develop an intuitive sense of structural action and reaction. You can find out **exactly** why symmetrical plans are good, as well as many ways of circumventing them if they do not suit your site, program or building image. The author's many years of experience teaching architectural students have enabled him to expand the range and refine the detail of his descriptions, and ensure their intelligibility.

Finally, if the architect still resists the effort to understand the earthquake, it must be remembered that we are not talking about an intellectual or aesthetic game, but knowledge and its application that may, in some future unknown event, save lives, reduce injuries and lessen economic and social catastrophe. Besides which, the whole subject is inherently fascinating.

Christopher Arnold

PREFACE

This book draws upon my structural engineering experience designing in the southern tip of the Pacific Rim of Fire, followed by twenty years teaching in a School of Architecture. Seismic design is a significant component in my Structures courses. These courses consist of formal lectures and tutorials, while including informal sessions where students are helped to develop seismic and gravity structure for their own architecture studio design projects. One of the most satisfying aspects of this less informal teaching is when students utilize structure not only to resist seismic and gravity forces but also to enrich their architectural design concepts.

The premise underlying this book is that structure is an indispensable architectural element imbued with the possibility of enhancing architectural functions and qualities. For example, appropriately designed structure can articulate entry into a building and celebrate interior circulation. It can create spaces and provide opportunities for aesthetic delight. So in the first instance, at preliminary design stage, structure needs to be designed by an architect.

The approach and content of the book is based upon that view of an architect's role in seismic design. If structure is to play architectural roles other than load-bearing, its design cannot be left to someone else. An architect should have the skills to conceive the structural configuration at the preliminary design stage that not only satisfies programmatic requirements and his or her design ideas, but is structurally sound especially with respect to seismic forces. Subsequent to this conception of structure, and ideally during that preliminary design process, structural engineering collaboration is indispensable. Ideally a structural engineer with specialist technical skills – and a sensitivity towards architectural aspirations – works alongside the architect to develop and refine the initial structural form. The engineer, designing well beyond the technical abilities of the architect then determines member sizes and attends to all the other structural details and issues.

Given the ideal situation outlined above, the book focuses on the core knowledge that architects require to 'outwit the quake'. Written for those designing buildings, its explanations provide the background, understanding, strategies and approaches to be applied in design.

Seismic principles and concepts rather than code requirements are emphasized. With a few exceptions, the book recognizes both the reality of architectural practice and architects' preferences by leaving equations and calculations to structural engineers.

The intended readership is primarily architectural students and architects – hence the generous number of explanatory diagrams and images, and the exclusion of civil engineering structures like bridges, wharfs and dams. However, the conceptual treatment of seismic resistance will also appeal to students of structural engineering and engineers who appreciate a non-mathematical introduction to seismic design. The qualitative approach herein complements engineers' more calculation-intensive analysis and design methods, and covers the design of components such as non-structural elements that most engineering texts and codes treat very briefly.

The chapter sequence of the book reflects a general progression in complexity. The gradual introduction of more complex issues is appropriate for architectural, architectural engineering and building science programmes. For example, the content of Chapters 1 and 2 is suited to first or second year courses, Chapters 3 to 5 to second or third years, and Chapters 6 to 11 to third or fourth years. Other chapters, especially Chapters 13 and 14 can be inserted into the senior years of a programme. The amount of material from the book that can be introduced into given courses may depend upon how much time a school's curriculum allocates to Structures. The non-mathematical approach of this book suggests a reappraisal of how Structures might be taught. If emphasis upon the quantitative treatment of Structures is reduced in favour of the introduction of a broader range of structural topics taught qualitatively, then space can be created for more material on seismic design.

Andrew Charleson

ACKNOWLEDGEMENTS

I am very grateful for help received during the preparation of this book. In particular I thank the following:

- Victoria University of Wellington for research and study leave to begin work on the book and for research grants for diagram preparation
- Professor Mary Comerio and the Visiting Scholar Program, Institute of Urban & Regional Development, University of California, Berkeley
- Those individuals and organizations that have provided images and granted permission for their use (unacknowledged images are by the author)
- Paul Hillier for photographic assistance
- Christopher Greenfield for drawing the diagrams
- The scientists, structural engineers and architects who each reviewed a chapter: Warwick Smith, Reagan Potangoroa (two chapters), Les Megget, David Whittaker, Win Clark, Alistair Cattanach, Brabha Brabhaharan, Peter Johnstone, Geoff Sidwell, Arthur Park, Peter Smith, Rob Jury, Guy Cleverley, Trevor Kelly, Bill Robinson, Jim Cousins, Graeme McIndoe, Geoff Thomas, Jitendra Bothara and Luke Allen. Randolph Langenbach commented on various sections of the manuscript, and
- My wife Annette for her support.

Finally, I acknowledge the use of Frank Lloyd Wright's phrase 'outwitting the quake' as the book's subtitle and in numerous occasions throughout the text. Following his insightful but ultimately flawed design of the Imperial Hotel, Tokyo that involved 'floating' the building on a deep layer of 'soft mud' in combination with a flexible superstructure, he writes: 'Why fight the quake? Why not sympathize with it and outwit it?' (Wright, F.L., 1977, *Frank Lloyd Wright: An Autobiography*. Quartet Books, Horizon Press, New York, p. 238).

1

EARTHQUAKES AND GROUND SHAKING

INTRODUCTION

According to the Natural History Museum, London, the ground upon which we build is anything but solid. The Earth Gallery illustrates how rocks flow, melt, shatter, are squeezed and folded. But more than that, the continents that support the earth's civilizations are in constant motion. Hundreds of millions of years ago the continents were joined, but now they are dispersing ever so slowly. Once, the east coast of South America nestled neatly against the west coast of Africa. Now, separated by the Atlantic Ocean, they lie 9600 km apart. The idea that buildings are founded upon stationary ground is an illusion. From the perspective of geological time, the earth's crust is in a state of dynamic flux.

The scientific understanding of this dynamic process known as continental drift or tectonic plate movement – the basic cause of most earthquakes – dates back only 100 years. Prior mythology and speculation that sought to explain earthquake occurrence and its prevention is deeply embedded in many cultures. For example, some peoples attributed earthquakes to subterranean beings holding up the world. Whether in the form of fish, animals or people, when they changed position to relieve their unrelenting burden, the earth shook. Many cultures possessed or still possess their own god or gods of earthquakes. Peoples like the Central Asian Turks valued jade as a talisman credited with the power to protect them from, among other dangers, earthquakes. Aristotle's influential belief was closer to the mark. It dismissed the activities of gods or other creatures in favour of natural phenomena. Namely, 'that mild earthquakes were caused by wind escaping from caves within the bowels of the earth and severe shocks by gales that found their way into great subterranean caverns.'[1]

It is not surprising that people sought to explain the occurrence of earthquakes, which happened without warning and so quickly devastated their communities. Although it appears that some animals, fish and insects sense and react to earthquakes before they are felt by humans, earthquakes strike suddenly. Often a rumbling is heard several seconds before shaking begins, and within a few seconds the initial tremors have grown into violent shaking. At other times a quake strikes like an instantaneous pulse. A reporter covering the October 2005 Pakistan earthquake recounts the experience of a Balakot boy searching through the rubble of his school where 400 of 500 of his fellow students had been buried alive. The boy recounted that the collapse occurred so suddenly, prompting the reporter to explain: 'How quick is hard to comprehend. At another school a teacher told a colleague of mine from the *Daily Telegraph* how he had just arrived at the door of his classroom. The children stood up. As they began their morning greeting of 'Good morning, Sir' the earthquake hit. The teacher stepped back in surprise, the roof collapsed. They all died, all 50 of them, just like that. No wobbling walls and dashes for the door. No warning. One second you have a classroom full of children in front of you, and the next, they are dead'.[2]

If the potential source of an earthquake attack is both known with reasonable confidence and is also some distance from a major city, an early warning system can be implemented. For instance, earthquakes most likely to damage Mexico City originate along the Guerrero coast some 280 km to the west. The 72 seconds that the earthquake waves take to travel to the city afford sufficient time for people to flee low-rise constructions or move to a safer location within their building. Commercial radio stations, the internet and audio alerting systems such as local sirens alert people to impending danger.[3] Several other cities, including Tokyo, have also installed early warning systems, but these allow far less time for preventative actions.[4] Unfortunately, for the vast majority of us living in seismic zones, any warning remains a dream.

Upon sensing initial ground or building movement, sufficient time usually elapses for the occupants to experience uncertainty and then fear. After realizing that the movement is not caused by a passing heavy vehicle but by an earthquake, one questions whether the vibrations are a precursor to more severe ground motion. While low-intensity earthquake shaking may be experienced as a gentle shock or small vibrations, during intense shaking people cannot walk steadily. They may be thrown over, or if sleeping, hurled out of bed. The perception of earthquake shaking is also usually heightened by what is happening in the immediate vicinity of the person experiencing a quake. Objects sliding, toppling or falling – be they building contents or elements of buildings such as suspended ceiling tiles, or dust from cracking plaster and concrete – all increase the psychological and physical trauma of a quake.

Apart from the poorest of communities for whom even partial earthquake protection is unaffordable, most of the disastrous effects of earthquakes are avoidable. Earthquake-resistant construction greatly reduces the loss of life from a damaging quake, as well as lessening economic losses and disruption to societal activities. Architects and structural engineers achieve earthquake-resistant buildings by following the principles and techniques outlined in this book. These are incorporated into new buildings with minor additional cost. The exact per centage increase in construction cost depends on many factors including the type and weight of building materials, the seismicity of the region and local code requirements. However, it is certainly far less expensive than improving the seismic performance of *existing* buildings.

Individuals, businesses and communities respond differently to the potential hazards posed by quakes. Although most earthquake-prone countries possess codes of practice that stipulate minimum standards of design and construction, particularly in developing countries, the majority of people are at considerable risk. Due to their economic situation or lack of appreciation of their seismic vulnerability, their homes and workplaces possess little if any seismic resistance. Every community in a seismically active zone should have numerous strategies to cope with a damaging quake. Some communities, due to their preoccupation with day-to-day survival, take a fatalistic approach that excludes any preventative or preparatory actions. Others implement civil defence and disaster management planning. Although not reducing the risk of injury or loss of life nor damage to buildings and infrastructure significantly, these initiatives reduce the trauma following a quake and assist post-earthquake restoration.

Quakes strike at the heart of a community. When they damage and destroy buildings, people and animals are injured and killed. Quakes destroy the basic necessities of life, demolishing shelter, ruining food and water supplies and disrupting people's livelihoods. Conversely, buildings that perform well during an earthquake limit its impact on people and their basic needs. The aim of this book is to reduce earthquake-induced devastation by providing architects and engineers with the knowledge to design both new and rehabilitated buildings that possess adequate seismic resistance.

UNDERSTANDING EARTHQUAKES

This section explains why architects might need to design earthquake-resistant buildings. It introduces the basic geological mechanisms causing earthquakes, explaining where and when earthquakes occur and the characteristics of ground shaking relevant to buildings. The focus here is upon those aspects of earthquakes over which we as designers have no control. Having outlined in this chapter what might be termed the earthquake problem, the remaining chapters deal with the solutions. For more detailed yet not too highly technical information on the basics of earthquake occurrence, the reader can refer to one of several general introductory texts.[5]

Why earthquakes occur

Compared to the 6400 km radius of the earth, the thickness of the earth's crust is perilously thin. The depth of the continental crust averages 35 km, and that of the oceanic crust only 7 km. While an analogy of the earth's crust as the cracked shell of a hen's egg exaggerates the thickness and solidity of the crust, it does convey the reality of a very thin and relatively brittle outer layer underlain by fluid − molten rock. Convection currents within the earth's viscous mantle, powered by vast amounts of thermal energy radiating from the earth's core, generate forces sufficiently large to move the continents. The earth's tectonic plates are like fragments of a cracked egg shell floating on fluid egg white and yolk. They move relative to each other approximately 50 mm per year; apparently about as fast as our fingernails grow (Fig. 1.1).

In some places, tectonic plates slip past each other horizontally. In others, such as where an oceanic plate pushes against a continental plate, the thinner oceanic plate bends and slides under the continental plate while raising it in a process known as *subduction* (Fig. 1.2). Due to the roughness of the surfaces and edges of tectonic plates, combined with the huge pressures involved, potential sliding and slipping movements generate friction forces large enough to lock-up surfaces in contact. Rather than sliding past each other, rock in a plate boundary area (say along a fault line) absorbs greater and greater compression and shear strains until it suddenly ruptures (Fig. 1.3). During rupture, all of the accumulated energy within the strained rock mass releases in a sudden violent movement − an earthquake.

The mechanical processes preceding an earthquake can be likened to the way we snap our fingers. We press finger against thumb to generate friction (Fig. 1.4(a)), then also using our finger muscles we apply

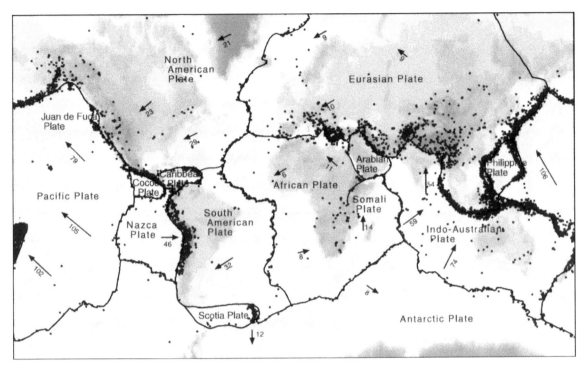

▲ 1.1   Tectonic plates and their annual movement (mm). The dots indicate positions of past earthquakes
(Reproduced with permission from IRIS Consortium).

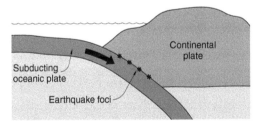

▲ 1.2   Subduction of an oceanic plate under a continental plate.

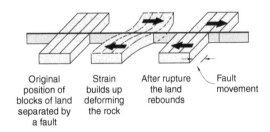

▲ 1.3   Increase of strain adjacent to a fault plane and the subsequent energy release and fault displacement.

a sideways force at the interface between the surfaces (Fig. 1.4(b)). If the initial pressure is low, they slide past each other without snapping. Increasing the pressure and the sideways force distorts the flesh. When the sliding force exceeds the friction between thumb and finger, the finger suddenly snaps past the thumb and strikes the wrist as the pent-up strain converts to kinetic energy (Fig. 1.4(c)).

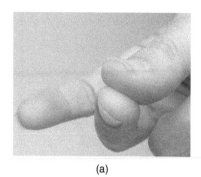
(a)

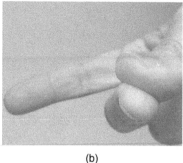
(b)

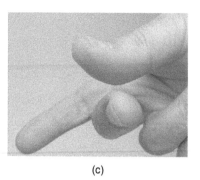
(c)

▲ **1.4** Experience the build-up of tectonic strain and energy release by snapping your fingers. Apply pressure normal to your finger and thumb (a), next apply sideways force (b), and then feel the sudden snapping when that force exceeds the friction between thumb and finger (c).

▲ **1.5** A surface fault with considerable vertical displacement. The 1999 Chi Chi, Taiwan earthquake.
(Reproduced with permission from Chris Graham).

The surface along which the crust of the earth fractures is an earthquake *fault*. In many earthquakes the fault is visible on the ground surface. Some combination of horizontal and vertical displacement is measurable, often in metres (Fig. 1.5). Chapter 15 discusses the wisdom of building over or close to active surface faults. The length of a fault is related to the earthquake magnitude (defined in a later section). For example, the fault length from a magnitude 6 quake is between 10–15 km, and 100–200 km long for a magnitude 8 event. The vertical dimension of a fault surface that contributes to the total area ruptured is also in the order of kilometers deep. The point on the fault surface area considered the centre of energy release is termed the *focus*, and its projection up to the earth's surface, a distance known as the *focal depth*, defines the *epicentre* (Fig. 1.6).

The length of the focal depth indicates the damage potential of an earthquake. Focal depths of damaging quakes can be several hundred kilometers deep. While perhaps not producing severe ground shaking, these deep-seated earthquakes affect a wide area. In contrast, shallower earthquakes concentrate their energy in epicentral regions. They are generally more devastating than deeper quakes where occurring near built-up areas. The focal depth of the devastating 2003 Bam, Iran earthquake that killed over 40,000 people out of a population of approximately 100,000, was only 7 km, while that of the similar magnitude 1994 Northridge, California quake was 18 km. The relatively low loss of life (57 fatalities) during the Northridge earthquake was attributable to both a greater focal depth, and more significantly, far less vulnerable building construction.

Epicentre — Surface faulting — Epicentral distance — Site of interest

Focal depth

Focus

Fault plane — Seismic waves

▲ **1.6** Illustration of basic earthquake terminology.

Where and when earthquakes strike

Relative movement between tectonic plates accounts for most continental or land-affecting earthquakes. Seventy per cent of these quakes occur around the perimeter of the Pacific plate, and 20 per cent along the southern edge of the Eurasian plate that passes through the Mediterranean to the Himalayas. The remaining 10 per cent, inexplicable in terms of simple tectonic plate theory, are dotted over the globe (Fig. 1.7). Some of these *intraplate* quakes, located well away from plate boundaries are very destructive.

A reasonably consistent pattern of annual world-wide occurrence of earthquakes has emerged over the years. Seismologists record many small but few large magnitude quakes. Each year about 200 magnitude 6, 20 magnitude 7 and one magnitude 8 earthquakes are expected. Their location, apart from the fact that the majority will occur around the Pacific plate, and their timing is unpredictable.

Although earthquake prediction continues to exercise many minds around the world, scientists have yet to develop methods to predict

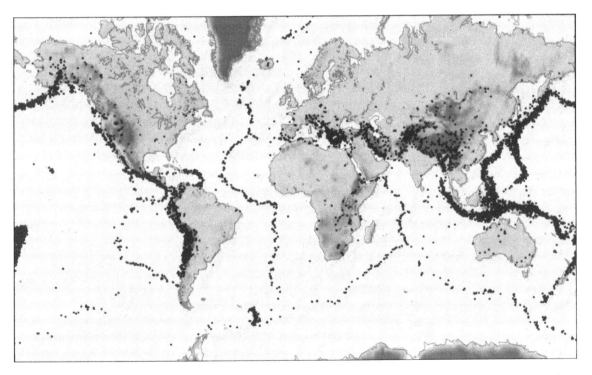

▲ **1.7**   Geographic distribution of earthquakes. Each dot on the map marks the location of a magnitude 4 or greater earthquake recorded over a period of five years.
(Reproduced with permission from IRIS Consortium).

precisely the location, time and magnitude of the next quake in a given geographic region. However, based upon a wide range of data including historical seismicity, measurements of ground uplift and other movement, and possible earthquake precursors such as foreshocks, scientists' predictions are more specific and refined than those of global annual seismicity discussed previously. The accuracy of such predictions will improve as seismological understanding continues to develop. Here are several examples of state-of-the-art predictions from peer reviewed research:

- 'There is a 62 per cent probability that at least one earthquake of magnitude 6.7 or greater will occur on a known or unknown San Francisco Bay region fault before 2032',[6]
- The probability of the central section of the New Zealand Alpine Fault rupturing in the next 20 years lies between 10 and 21 per cent,[7] and
- The probability of Istanbul being damaged by an earthquake greater or equal to magnitude 7 during the next thirty years is 41 ± 14 per cent.[8]

Several other valid generic predictions regarding quakes can be made; a large quake will be followed by *aftershocks*, a quake above a given magnitude event is implausible within a given geographic region, and certain size quakes have certain recurrence intervals.

In the hours and even months following a moderate to large earthquake, aftershocks or small earthquakes continue to shake the affected region. Although their intensities diminish with time, they cause additional damage to buildings weakened by the main shock, like the magnitude 5.5 aftershock that occurred a week after the 1994 Northridge earthquake. Post-earthquake reconnaissance and rescue activities in and around damaged buildings must acknowledge and mitigate the risks aftershocks pose.

Some predictions, such as a region's *maximum credible earthquake*, are incorporated into documents like seismic design codes. Based mainly upon geological evidence, scientists are confident enough to predict the maximum sized quake capable of occurring in a given region. For example, the largest earthquake capable of being generated by California's tectonic setting is considered to be magnitude 8.5. Its *return period*, or the average time period between recurrences of such huge earthquakes is assessed as greater than 2500 years.

Structural engineers regularly use predicted values of ground accelerations of earthquakes with certain return periods for design purposes. The trend is increasing for seismic design codes to describe the design-level earthquake for buildings in terms of an earthquake with a certain average return period. This earthquake, for which even partial building collapse is unacceptable, is typically defined as having a 10 per cent

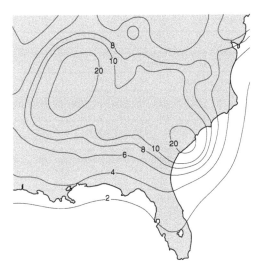

probability of being exceeded within the life of a building, say 50 years. The return period of this design earthquake is therefore approximately 500 years.

The probability p of an earthquake with a given return period T occurring within the life of a building L can be calculated using Poisson's equation, $p = 1 - e^{-L/T}$. For example, if L = 50 years, and T = 500 years, the probability of this event being exceeded during the lifetime of the building is approximately 0.1 or 10 per cent.

Special buildings that require enhanced seismic performance, like hospitals and fire stations, are designed for larger quakes. In such cases design earthquake return periods are increased, say to 1000 or more years. Designers of these important buildings therefore adopt higher design acceleration values; the longer the return period, the larger the earthquake and the greater its ground accelerations. Figure 1.8 shows a portion of a typical seismic map.[9] Most countries publish similar maps.

EARTHQUAKE MAGNITUDE AND INTENSITY

Seismologists determine the position of a quake's epicentre and its *magnitude*, which relates to the amount of energy released, from seismograph records. The magnitude of a quake as determined by the *Richter Scale* relates logarithmically to the amount of energy released. An increase of one step in magnitude corresponds to an approximate 30-fold increase in energy, and two steps, nine hundred times more energy. The 1976 Tangshan earthquake in China, the twentieth century's most lethal earthquake that caused approximately 650,000 fatalities, was magnitude 7.7.[10] The largest ever recorded quake was the magnitude 9.5 in the 1960 Great Chilean earthquake which, even with its devastating tsunami, had a significantly lower death toll. So the value of magnitude itself does not indicate the impact of a quake. Large earthquakes in regions distant from built-up areas may pass almost unnoticed. Another form of measurement describes the degree of seismic damage a locality suffers or is likely to suffer.

While each earthquake is assigned a single magnitude value, the *intensity* of earthquake shaking varies according to where it is felt. A number of factors that include the earthquake magnitude, the distance of the site from the epicentre, or *epicentral distance* (see Fig. 1.6) and the local soil

▼ **1.1**   Partial summary of the Modified Mercalli Intensity (MMI) Scale

Intensity	Description
I to III	Not felt, except under special circumstances.
IV	Generally felt, but not causing damage.
V	Felt by nearly everyone. Some crockery broken or items overturned. Some cracked plaster.
VI	Felt by all. Some heavy furniture moved. Some fallen plaster or damaged chimneys.
VII	Negligible damage in well designed and constructed buildings through to considerable damage in construction of poor quality. Some chimneys broken.
VIII	Depending on the quality of design and construction, damage ranges from slight through to partial collapse. Chimneys, monuments and walls fall.
IX	Well designed structures damaged and permanently racked. Partial collapses and buildings shifted off their foundations.
X	Some well-built wooden structures destroyed along with most masonry and frame structures.
XI	Few, if any masonry structures remain standing.
XII	Most construction severely damaged or destroyed.

▲ **1.9**   A map showing the distribution of Modified Mercalli Intensity for the 1989 Loma Prieta, California earthquake. Roman numerals represent the intensity level between isoseismal lines, while numbers indicate observed intensity values.
(Adapted from Shephard et al., 1990).

conditions influence the intensity of shaking at a particular site. An earthquake generally causes the most severe ground shaking at the epicentre. As the epicentral distance increases the energy of seismic waves arriving at that distant site as indicated by the intensity of shaking, diminishes. Soft soils that increase the duration of shaking as compared to rock also increase the intensity. One earthquake produces many values of intensity.

Another difference between the magnitude of an earthquake and its intensities is that, whereas the magnitude is calculated from seismograph recordings, intensity is somewhat subjective. Intensity values reflect how people experienced the shaking as well as the degree of damage caused. Although several different intensity scales have been customized to the conditions of particular countries they are similar to the internationally recognized Modified Mercalli Intensity Scale, summarized in Table 1.1. Based on interviews with earthquake survivors and observations of damage, contours of intensity or an *isoseismal* map of an affected region, can be drawn (Fig. 1.9).[11] This information is useful for future earthquake studies. It illustrates the extent, if any, of an earthquake's directivity, how the degree of damage varies over a region with increasing epicentral distance, and how areas of soft soil cause increased damage.

THE NATURE OF EARTHQUAKE SHAKING

At the instant of fault rupture, seismic waves radiate in all directions from the focus. Like the waves emanating from a stone dropped into a pond, seismic waves disperse through the surrounding rock, although at far greater velocities. But unlike the ever increasing circles of pond waves, the spread of seismic waves can take more elliptical forms. In these situations where the earthquake energy partially focuses along one certain direction, the earthquake exhibits *directivity*. The extent of directivity, which causes more intense damage over the narrower band in the line of fire as it were, is unpredictable. Directivity depends on several geological factors including the speed at which the fault rupture propagates along its length.

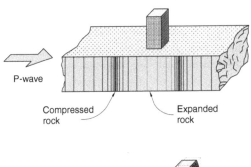

P-wave

Compressed rock

Expanded rock

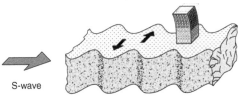

S-wave

▲ **1.10** Dynamic ground movements caused by the propagation of P- and horizontal S-waves.

Of the three types of waves generated by fault rupture, two travel underground through rock and soil while the third is confined to the ground surface. *P-waves*, or Primary waves, travel the fastest. They move through rock in the same way as sound waves move through air, or as a shock wave travels along a metal rod when it is struck at one end. They push and pull the soil through which they pass. *S-waves* or Shear waves, of most concern to buildings, move soil particles side to side, vertically and horizontally (Fig. 1.10). They propagate from the focus at a speed of about 3 km/sec. *Surface waves* are the third type of waves. Named after the scientists who discovered them, *Love* waves vibrate only in the horizontal plane on the earth's surface while *Rayleigh* waves also have a significant vertical component. Their up-and-down motion is similar to ocean waves. The author vividly recalls the peaks and troughs of Rayleigh waves travelling along the road when once, as a boy, he was riding to school.

Horizontal S-waves, Love and Rayleigh waves, all of which move the ground to-and-fro sideways, cause the most damage to buildings. Buildings are far more susceptible to horizontal rather than vertical accelerations. The snake-like action of these waves induces into the foundations of buildings horizontal accelerations that the superstructures then amplify. The waves also transmit horizontal torsion rotations into building foundations. The primary focus of seismic resistant design is to withstand the potentially destructive effects of these waves.

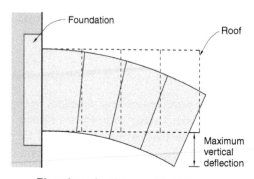

Elevation of a shear wall building

▲ **1.11**   A building tipped onto its side and cantilevered from its base experiences 1.0 g acceleration acting vertically.

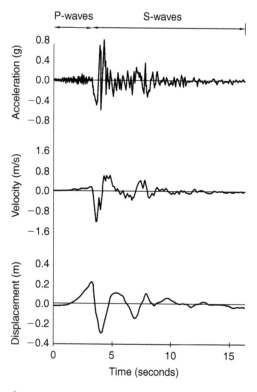

▲ **1.12**   North-south components of acceleration, velocity and displacement histories from Sylmar, California, during the 1994 Northridge earthquake. (Adapted from Norton *et al.*, 1994).

Characteristics of ground shaking

From the perspective of designing seismic resistant buildings, the three most important characteristics of ground shaking are the *value of peak ground acceleration*, the *duration of strong shaking* and the *frequency content of the shaking*. Recorded peak ground accelerations of damaging earthquakes range from 0.2g to over 1.0g where g is the acceleration due to gravity. A 1.0g horizontal acceleration at the base of a rigid building induces the same force as if the building were tipped onto its side to cantilever horizontally from its base (Fig. 1.11). Very few buildings can survive such a large force. The higher the level of ground acceleration, the greater the horizontal earthquake forces induced within the building. As explained in Chapter 2, the horizontal flexibility of the super-structure of a building amplifies the ground shaking commonly by a factor of up to two to three times.

Earthquake acceleration records are obtained from *seismographs* which record the rapidly changing accelerations or velocities throughout the duration of a quake. Mathematical manipulation of these records produces corresponding graphs of velocity and displacement against time (Fig. 1.12).[12] Ground motions are easiest to visualize from the graph of displacement against time. Figure 1.12 shows a movement of 0.2 m in one direction and just over 0.3 m in the other in a period of approximately 1.5 seconds. An appreciation of the maximum inertia forces generated within buildings during this quake is gained from noting the far higher frequency accelerations from which the peak ground acceleration can be determined. The accelerations last for such small periods of time their displacements are smoothed out in the displacement-against-time graph.

The *duration* of strong shaking also affects the degree of earthquake damage a building sustains. Just as a losing boxer, reeling from blow after blow to the body desperately awaits the end of the round, so a building is concerned about the duration of a quake. The longer shaking feeds dynamic energy into a building, causing more and more energy to be absorbed by the structure, while the extent of damage and its severity grows. In conventional reinforced concrete construction, once beams and columns crack, further load cycles cause the concrete on either side of cracks to be ground away, both weakening the structure and making it more flexible.

The duration of strong shaking correlates with earthquake magnitude and soil type.[13] Duration increases with magnitude. For a magnitude 6 earthquake, expect approximately 12 seconds of strong shaking, but the duration of a magnitude 8 quake increases to over 30 seconds. If a site is underlain by soil rather than rock, the duration of strong shaking doubles.

The frequency content of earthquake shaking at a given site is also significantly affected by the ground conditions. On a rock site, most of the earthquake energy is contained within frequencies of between 1 and 6 cycles per second. In contrast, soil sites reduce the frequency of high energy vibrations. As discussed in Chapter 2, the degree to which a building superstructure amplifies ground motions – and consequently requires enhanced seismic resistance – depends on how close the frequencies of energy-filled vibrations match the natural frequency of the building.

Another important characteristic of ground shaking is its *random directivity*. Even though the predominant shaking of a quake may be stronger in one particular direction, for design purposes ground shaking should always be considered totally random in three dimensions. Figure 1.13 shows an example typical of the chaotic and irregular movements caused by earthquakes. Random directional shaking has major consequences for earthquake resistant buildings. As discussed in the Chapter 2, buildings must be designed for earthquake forces acting in *any* direction.

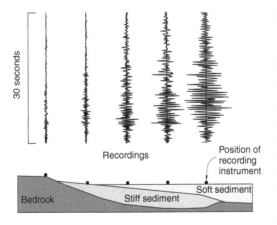

▲ 1.13 A scratch plate accelerometer record of a small earthquake. It shows directionally-random horizontal accelerations. The numbered rings indicate acceleration values expressed as a decimal of the acceleration due to gravity. (Reproduced with permission from GNS Science).

30 seconds

Recordings

Position of recording instrument

Soft sediment

Bedrock   Stiff sediment

▲ 1.14 A cross-section through a geological setting near Wellington, showing acceleration records at five sites during a small earthquake. Note how the accelerations increase and frequencies reduce above deeper and soft sediments. (Reproduced with permission from J. Taber).

IMPORTANCE OF GROUND CONDITIONS

The influence of soil in reducing the frequency of ground shaking measured in cycles per second while increasing its duration and severity has been mentioned. Local soil conditions, particularly deep layers of soft soil as may be found in river valleys or near estuaries, significantly amplify shaking. They also modify the frequency content of seismic waves by filtering out higher frequency excitations (Fig. 1.14). Although this effect is observed in many quakes it was particularly evident in a local area of Mexico City during the 1985 Mexico earthquake. A small area of the city built over a former lake bed is underlain by deep soft clay. During the earthquake this soft soil deposit behaved like a bowl of soft jelly shaken by hand. The soil amplified the vibrations of the seismic waves in the bedrock at the base of the soft soil by factors greater than five times and shook to-and-fro with a natural frequency of 0.5 cycles per second. This shaking, considerably slower than that measured on bedrock nearby, caused modern high-rise buildings with similar natural frequencies to resonate. Some collapsed, and many were badly damaged.

REFERENCES AND NOTES

1   Levy, M. and Salvadori, M. (1995). *Why the Earth Quakes: the story of earthquakes and volcanoes*. W.W. Norton & Company, Inc.
2   Danahar, P. (2005). Class cut short. *Outlook India*, 31 October, 80.
3   Espinso-Arnda, J.M. and Rodrigues, F.H. (2003). The seismic alert system of Mexico City. In *International Handbook of Earthquake and Engineering Seismology*, Lee, W.H.K. et al. (eds.). Academic Press, Vol. 81B, pp. 1253–1259.
4   Erdik, M. Urban earthquake rapid response and early warning systems, *First European Conference on Earthquake Engineering and Seismology*, Geneva, Switzerland, 3–8 September, Paper K4.
5   For a comprehensive and accessible introduction refer to Bolt, B.A. (2004). *Earthquakes* (5th Ed.). W.H. Freeman & Co.
6   US Geological Survey (2006). *Putting Down Roots in Earthquake Country – your handbook for the San Francisco Bay region*. United States Government Printing Office, p. 7. Available at www.pubs.usgs.gov/gip/2005/15/.
7   Rhoades, D.A. and Van Dissen, R.J. (2003). Estimates of the time-varying hazard of rupture of the Alpine Fault, New Zealand, allowing for uncertainties. *New Zealand Journal of Geology & Geophysics*, **46**, 479–488.
8   Parsons, T. (2004). Recalculated probability of M ≥ 7 earthquakes beneath the Sea of Marmara, Turkey, *Journal of Geophysical Research*, **109**, B05304, pp. 21.
9   Adapted from US Geological Survey (1996). Central and Eastern US hazard maps. From http://earthquake.usgs.gov/research/hazmaps.
10 Huixian, L. et al. (2002). *The Great Tangshan Earthquake of 1976*, Technical Report: Caltech EERL:2002:001, California Institute of Technology. Available at http://caltecheerl.library.caltech.edu/353/.
11 Shephard, R.B. et al. (1990). The Loma Prieta, California, earthquake of October 17, 1989: report of the NZNSEE reconnaissance team. *Bulletin of the New Zealand National Society for Earthquake Engineering*, **23**:1, 1–78.
12 Norton, J.A. et al. (1994). Northridge earthquake reconnaissance report. *Bulletin of the New Zealand National Society for Earthquake Engineering*, **27**:4, 235–344.
13 Mohraz, B. and Sadek, F. (2001). Chapter 2, Earthquake ground motion and response spectra. In *The Seismic Design Handbook* (2nd Edn), Naeim, F. (ed.). Kluwer Academic Publishers, 47–124.

2

How buildings resist earthquakes

Introduction

Chapter 1 dwelt with the nature of ground shaking as it affects buildings. This chapter now outlines the basic principles of seismic resistance for buildings. Factors such as the dynamic characteristics of earthquakes, their duration and the effects of site conditions are all external to a building. No matter how well or poorly designed, a building has no control over those effects. But as we shall see, a combination of factors such as the form of a building, its materials of construction and dynamic characteristics, as well as the quality of its structural design and construction, greatly influence how a building responds to any shaking it experiences.

We therefore turn our attention to those aspects of a building itself that largely determine its seismic response. This chapter begins by discussing the nature of earthquake forces and notes how they differ from other forces such as those caused by the wind, that also act upon buildings. The following sections then explore the key physical properties that affect the severity of seismic forces. After appreciating those factors that influence levels of seismic force, the basic requirements for seismic resistance are considered. This in turn leads to an introduction to building torsion and the concept of force paths.

Nature of seismic forces

Seismic forces are inertia forces. When any object, such as a building, experiences acceleration, inertia force is generated when its mass resists the acceleration. We experience inertia forces while travelling. Especially when standing in a bus or train, any changes in speed (accelerations) cause us to lose our balance and either force us to change our stance or to hold on more firmly.

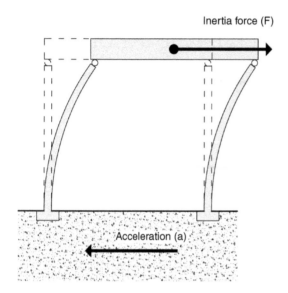

Inertia force (F)

Acceleration (a)

▲ **2.1**   An inertia force is induced when a building (with cantilever columns) experiences acceleration at its base.

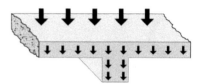

Gravity loads and forces

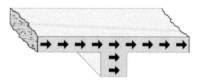

Horizontal inertia forces

▲ **2.2**   An area of concrete floor showing the difference between gravity forces and horizontal inertia forces.

Newton's Second Law of Motion, $F = M \times a$ enables the inertia force F to be quantified. M, the mass of an object, is determined by dividing its weight by the acceleration due to gravity, while a is the acceleration it is subject to (Fig. 2.1). This is *the* primary equation for seismic resistant design.

Inertia forces act *within* a building. They are internal forces. As the ground under a building shakes sideways, horizontal accelerations transfer up through the superstructure of the building and generate inertia forces throughout it. Inertia forces act on every item and every component. Every square metre of construction, like a floor slab or wall, possesses weight and therefore mass. Just as gravity force that acts vertically is distributed over elements like floor slabs, so is seismic inertia force, except that it acts horizontally (Fig. 2.2).

The analogy between gravity and inertia forces can be taken further. As the sum of gravity forces acting on an element can be assumed to act at its centre of mass (CoM), so can the inertia force on any item be considered to act at the same point. Since most of the weight in buildings is concentrated in their roofs and floors, for the sake of simplicity designers assume inertia forces act at the CoM of the roof and each floor level (Fig. 2.3). For most buildings the CoM corresponds to the centre of plan.

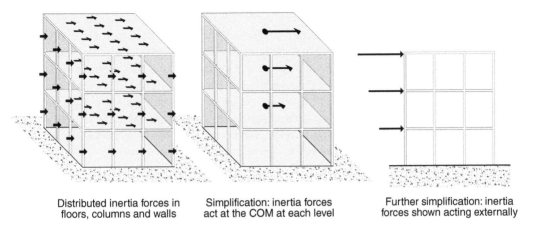

Distributed inertia forces in floors, columns and walls

Simplification: inertia forces act at the COM at each level

Further simplification: inertia forces shown acting externally

▲ **2.3** Increasing simplification of how inertia forces on a building are expressed graphically.

At this point a significant difference between wind and inertia forces can be appreciated. Wind force is *external* to a building. Wind pressure that pushes against a building acts upon external surfaces. Its magnitude and centre of loading is determined by the surface area upon which it acts (Fig. 2.4). Like inertia forces, wind loading is dynamic, but whereas peak earthquake forces act for just fractions of a second, the duration of a strong wind gust lasts in the order of several seconds. Another difference between the two load conditions is that inertia

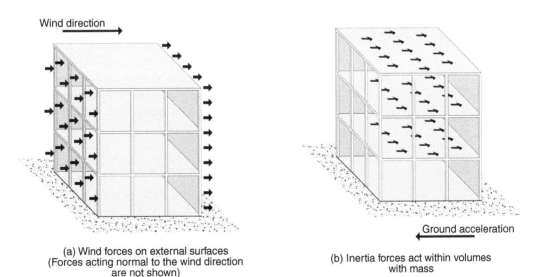

(a) Wind forces on external surfaces
(Forces acting normal to the wind direction are not shown)

(b) Inertia forces act within volumes with mass

▲ **2.4** Comparison between externally acting wind forces and internal inertia forces.

forces are cyclic – they act to-and-fro. In spite of these significant differences the feature common to both forces is that they act horizontally. Although near-vertical wind suction forces act on roofs during
a wind storm and vertical ground accelerations also occur during an
earthquake, these vertical forces usually have little impact on the overall behaviour of buildings. The only time a building might need to be
explicitly designed for vertical accelerations is where it incorporates
some long-spanning floor or roof structures, say in excess of 20 m
length, or significant horizontal cantilevers.

Factors affecting the severity of seismic forces

Building weight

The single most important factor determining the inertia force in a
building is its weight. Newton's Law states that inertia force is proportional to mass or weight. The heavier an object the greater the inertia
force for a certain level of acceleration. In earthquake prone regions,
we should therefore build as light-weight as practicable to reduce seismic vulnerability. Wherever possible, lighter elements of construction
should be substituted for and replace those that are heavier.

Unfortunately, in most countries common and economical forms of
construction are heavy. Brick or stone masonry, adobe and reinforced
concrete are the most widely used materials. In those areas where
wood is still plentiful light-weight wood framed construction is an
option, but the reality for most people is to inhabit heavy buildings.
Nevertheless, architects and structural engineers should always
attempt to build more lightly, bearing in mind economy and other
factors like sustainability. This intent is applicable for both new buildings and those being renovated or retrofitted. There are often opportunities to reduce building weight by, for example, demolishing heavy
interior masonry walls and replacing them with light timber or steel
framed construction.

Natural period of vibration

Hold a reasonably flexible architectural model of a building and give
it a sharp horizontal push at roof level. The building will vibrate back
and forth with a constant period of vibration. As illustrated in Fig. 2.5,
the time taken for one full cycle is called the *natural period of vibration*,
measured in seconds. Every model and full-scale building has a natural
period of vibration corresponding to what is termed the *first mode of
vibration*. Depending on the height of a building there may be other

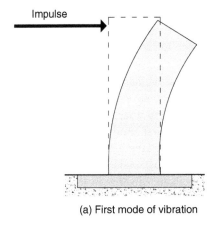

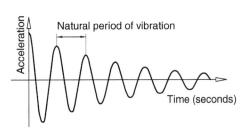

(a) First mode of vibration

(b) A record of the building acceleration after the impulse

▲ **2.5** A building given an impulsive force (a) and subsequent vibrations at its natural period of vibration (b).

periods of vibration as well. They correspond to the second, third and higher modes of vibration (Fig. 2.6(a)). There are as many modes of vibration as there are storeys in a building. But usually the effects of the first few modes of vibration only need to be considered by a structural engineer. Higher modes that resonate less strongly with earthquake shaking contain less dynamic energy.

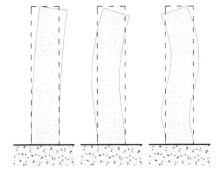

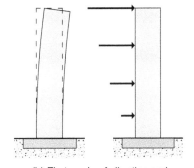

(a) First three modes of vibration of a vertical tower

(b) First mode of vibration and corresponding inertia forces

▲ **2.6** The deflected shapes of the first three modes of vibration (a) and the first mode of vibration as the source of most inertia force (b).

When earthquake waves with their chaotic period content strike the foundations of a building, its superstructure responds to the various periods of vibration that are all mixed-up together to comprise the shaking. The different periods of vibration embedded within the earthquake record cause corresponding modes of vibration in the building to resonate simultaneously. At any instant in time the deflected shape of a building is defined by the addition of many modes of vibration.

Particularly in low- to medium-rise buildings, most of the dynamic energy transmitted into them resonates the first mode and its natural period of vibration; and to a far lesser extent the second and higher modes. Because in the first mode every part of a building moves in the same direction simultaneously resulting in the greatest overall inertia force, it is the most important. Its mode shape, rather like an inverted triangle, explains why inertia forces acting at each floor level increase with height (Fig. 2.6(b)). Although the higher modes of vibration do not significantly affect the total inertia force to be resisted by the building at its base, they can cause very high 'whiplash' accelerations near the roof of a building. These localized yet intense horizontal accelerations often cause of increased damage to non-structural elements in upper storeys (Chapters 10 and 11).

The natural period of vibration of a building depends upon a number of factors:

- *Building height has the greatest influence.* The higher a building, the longer its natural period of vibration. A very approximate rule-of-thumb method for calculating the natural period of vibration is to multiply the number of storeys of a building by 0.1. The natural period of a ten-storey building is therefore approximately 1.0 second.
- *The weight of the building.* The heavier a building, the longer the natural period, and finally,
- *The type of structural system provided to resist seismic forces.* The more flexible or less stiff a structure, the longer its natural period. A moment frame structure, for example, is usually more flexible than a shear wall structure, so its natural period is longer.

In practice, natural periods of vibration vary between say 0.05 seconds for a stiff single-storey building to a period of approximately seven seconds for one of the world's tallest buildings at 101 storeys (Fig. 2.7).

▲ **2.7** One of the tallest buildings in the world, Taipei 101, Taiwan.

Damping

Damping is another important but less critical dynamic characteristic of a building. Fig. 2.5(b) illustrates how damping reduces the magnitude of horizontal vibrations with each successive cycle. Damping, mainly caused by internal friction within building elements, causes the amplitude of vibrations to decay. The degree of damping in a building depends upon the material of its seismic resisting structure as well as its other construction materials and details. Once the choice of materials has been made, the damping in a building to which its seismic response is reasonably sensitive, is established. Reinforced concrete structures possess more damping than steel structures, but less than those constructed of wood. However, the choice of structural materials is rarely if ever made on the basis of their damping values. Damping absorbs earthquake energy and reduces resonance or the build-up of earthquake inertia forces so it is very beneficial.

Without being aware of it, we regularly experience damping in cars. Shock-absorbers quickly dampen out vertical vibrations caused when a car rides over a bump on the road. Damping in buildings has the same but much smaller effect. Apart from high-tech buildings that might have specially designed dampers incorporated into their structural systems (Chapter 14), structural engineers do not intentionally attempt to increase damping. They just accept it and allow for its beneficial presence in their calculations. If the damping in a typical reinforced concrete building is halved, seismic response (peak acceleration) increases by approximately 30 per cent.

Response spectrum

The response spectrum is a convenient method for illustrating and quantifying how the natural period of vibration and damping of a building affects its response to earthquake shaking.

As schematically illustrated in Fig. 2.8 a digitally recorded earthquake accelerogram is the input signal to a dynamic hydraulic ram attached to

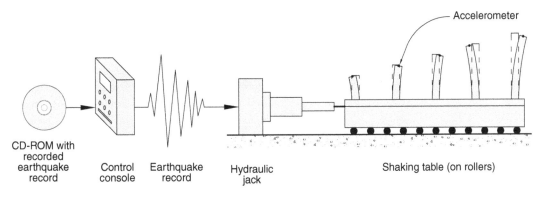

CD-ROM with recorded earthquake record

Control console

Earthquake record

Hydraulic jack

Shaking table (on rollers)

Accelerometer

▲ **2.8** Generating a response spectrum from an earthquake record using a shaking table.

a shaking table. Model buildings, each with a longer period of vibration from left to right, are mounted on the table, and an accelerometer is attached to the roof of each to measure its maximum horizontal acceleration. The buildings possess identical amounts of damping. When the shaking table simulates a recorded earthquake each building vibrates differently and its maximum acceleration is recorded and then plotted on a graph (Fig. 2.9(a)). Although the procedure outlined above using mechanical equipment like a shaking table could be used in practice, it is far more convenient to model the whole process by computer. All response spectra are computer generated.

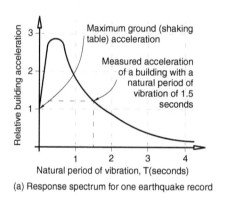

(a) Response spectrum for one earthquake record

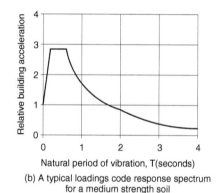

(b) A typical loadings code response spectrum for a medium strength soil

▲ **2.9** A typical response spectrum (a) and its expression in an earthquake loadings code (b).

The shape of a response spectrum illustrates how the natural period of vibration of a building has a huge effect on the maximum horizontal acceleration experienced, and consequently upon the magnitude of inertia force it should be designed for. With reference to Fig. 2.9(b), the maximum acceleration of a building with a natural period of 0.0 seconds is represented by 1.0 unit of acceleration. This point on the spectrum represents the *peak ground acceleration*. Buildings with certain longer natural periods of vibration amplify ground accelerations. For example, buildings with T = 0.2 to 0.7 seconds resonate with the cyclic ground accelerations, amplifying them by almost a factor of 3.0. As natural periods become longer, from 0.7 to 1.7 seconds, peak building accelerations reduce towards the same intensity as the peak ground acceleration. Beyond 1.7 seconds the maximum accelerations continue to diminish until at T = 4.0 seconds the building acceleration is only 0.3 of the maximum ground acceleration. So, depending on the value of the natural period of vibration an approximately ten-fold variation in maximum building acceleration is possible! A building with T = 4.0 seconds (approximately 40 storeys high) need be designed for only 10 per cent of the design force of a building of the same weight with T = 0.2 seconds (two storeys). In general, the longer the natural period of vibration, the less the maximum acceleration and seismic design force. Seismic isolation (Chapter 14) is little more than an application of this principle.

Although the shape of a particular response spectrum illustrates some of the fundamentals of seismic design it is not particularly useful for structural engineers. Ideally they need similar graphs for *future* damaging earthquakes. Then once they have calculated the natural period of vibration of a building they can determine its maximum acceleration, calculate inertia forces and then design the seismic resisting structure accordingly. To meet this need the best that earthquake engineers can do is to select a suite of past earthquake records as a basis for extrapolating into the future. Response spectra are generated and then averaged to obtain a design response spectrum that is included in a country's earthquake loading code (Fig. 2.9(b)). Earthquake recordings from different soil conditions account for how soil modifies bedrock shaking as discussed in the previous chapter. Most loadings codes provide four response spectra to represent rock sites and firm, medium and soft soil sites.

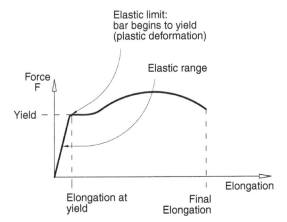

▲ **2.10** A graph of tensile force against elongation of a steel rod.

Ductility

Ductility has a large influence upon the magnitude of accelerations and seismic forces a building is designed for, just like its natural period of vibration. Depending upon the degree of ductility a structure possesses the design seismic force can be reduced to approximately as little as one sixth of an equivalent non-ductile structure.

So what is ductility? Think of it as the opposite of brittleness. When a brittle or non-ductile material like glass or concrete is stretched it suddenly snaps on reaching its elastic limit. A ductile material on the other hand like steel, reaches its elastic limit and then deforms plastically. It even slightly increases in strength until at a relatively large elongation it breaks (Fig. 2.10). Ductile (and brittle) performance, possible for all the

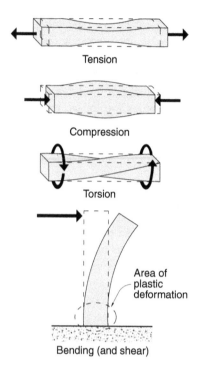

Tension

Compression

Torsion

Area of plastic deformation

Bending (and shear)

▲ **2.11** Different structural actions causing ductile deformations in structural elements.

structural actions illustrated in Fig. 2.11, can be easily demonstrated. Take 400 mm lengths of 3 mm diameter steel wire and 5×20 mm wood. Hold the wooden member vertically and firmly at its base and apply a horizontal force at its top. The wood suddenly snaps due to bending at its base. However, as the horizontal force at the top of a steel wire increases the steel at its base region yields in a ductile fashion. A *plastic hinge* or *structural fuse* forms where the bending moment exceeds the bending strength of the wire.[1] Plastic deformation occurs but the wire maintains its bending strength even though it has suffered permanent deformation. It requires just as much force to bend the wire back to its original position.

Ductile structural materials don't necessarily guarantee ductile structures. The critical cross-sections of members and their connections need to be properly proportioned and detailed to completely exploit the ductile nature of the material. For example, if a steel compression member is too long it suffers non-ductile buckling before being squashed plastically – a ductile overload mechanism. If the bolts or welds in its end connections are weaker than the member itself they break prematurely before the steel member yields in a ductile fashion.

Ductility is one of the most desirable structural qualities of seismic resisting structures. If the intensity of earthquake shaking exceeds the strength of a brittle member – be it a beam or column – the member breaks suddenly, possibly leading to building collapse. But if the member is ductile, its material will yield, exhibiting plastic behaviour up to a relatively large deflection. In the process of being deformed plastically, a ductile member absorbs seismic energy that would otherwise lead to the building experiencing increased accelerations. Ductility therefore increases the effective level of damping in a building.

The primary advantage of ductile members is their ability to form 'structural fuses'. Unlike electrical fuses which – depending on their era of construction – either blow a fuse wire or break a circuit, a structural fuse does not break or need resetting. A localized area of a structural member is merely stretched plastically. This deformation leads to damage but the fuse area or region is designed not to lose strength. In the process of fusing it prevents any more force entering the member or structure and causing damage elsewhere. See Chapter 3 for more on this.

Non-ductile buildings are designed for up to six times the force of those that are ductile. Because a non-ductile structure breaks in an overload situation it must be strong enough to resist the maximum anticipated inertia forces. The consequences of overload on a ductile structure are far less severe. Nothing snaps and although structural fuse regions suffer some damage, because they maintain their strength they prevent building collapse.

To some, the thought of ductile structures designed only for a mere fraction of the inertia force that would occur if the structure were to remain elastic, seems very non-conservative. Their concern would be valid if seismic forces were not cyclic nor characterized by short periods of vibration. It would be disastrous, for example, to design for only one sixth of the gravity forces acting on a structure; the structure would collapse. But because of the to-and-fro nature of earthquake shaking, and the fact that peak inertia forces in one direction act for less than half of a building's natural period of vibration – often less than one second – the approach of designing ductile structures for reduced forces is sound and is the basis of modern seismic loading codes.

RESISTING SEISMIC FORCES

To resist horizontal seismic forces successfully buildings must possess strength and stiffness, and in most cases ductility as well. Before getting into the detail covered by following chapters this section considers the structural necessities of strength and stiffness.

Strength

The superstructure of every building requires sufficient structural strength to resist the bending moments and shear forces[2] caused by seismic forces, and a foundation system capable of preventing overturning and sliding.

Consider the building shown in Fig. 2.12. Two shear walls resist inertia forces in both the x and y directions and transfer them to the foundations. The walls are subject to bending moments and shear forces for which they must be designed in order to satisfy the requirements of the seismic design code. Bending and shear actions, which increase from the roof level to reach their maximum values at the bases of the walls, are resisted by the foundations and transferred into the ground.

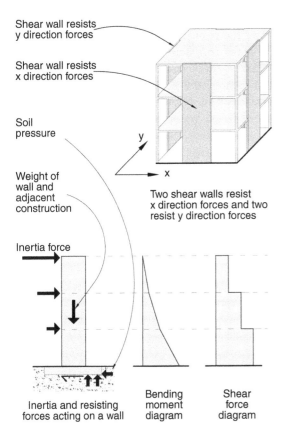

Shear wall resists
y direction forces

Shear wall resists
x direction forces

Soil
pressure

Weight of
wall and
adjacent
construction

Inertia force

Two shear walls resist
x direction forces and two
resist y direction forces

Inertia and resisting
forces acting on a wall

Bending
moment
diagram

Shear
force
diagram

▲ **2.12** A building with shear walls resisting inertia forces in both orthogonal directions and the wall forces, bending moment and shear force diagrams.

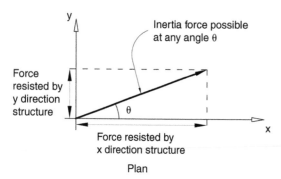

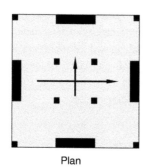

Plan

Relative magnitude of forces acting in plan in each direction for the force at an angle θ above

▲ **2.13** With strength in two orthogonal directions structure can resist earthquake attack from any direction. The building plan is from Fig. 2.12.

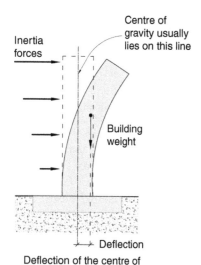

Deflection of the centre of gravity of the building

▲ **2.14** The combination of horizontal deflection and building weight increases the risk of toppling.

Due to the alignment of the shear walls which are strong only in the direction of their lengths, horizontal strength is provided in *both* the *x* and *y* directions. This provision of bi-directional strength responds to the fact that (as mentioned in Chapter 1) earthquake shaking is directionally random. Structure must be prepared for an earthquake attack from *any* direction. So long as strength is provided in any two orthogonal directions then *any* angle of attack is covered. A seismic force can be resolved into two orthogonal components which are resisted by structure with strength parallel to those directions (Fig. 2.13).

In a similar way as *x* and *y* direction structure resist seismic forces from any direction, structure not parallel to either the *x* or *y* axis provides strength along both axes. If the inertia force in Fig. 2.13 is considered to represent the strength of say a shear wall, then that wall contributes considerable strength in the *x* direction and less in the *y* direction. Refer to Figure 8.22 which shows how the strengths of non-orthogonal walls are resolved into *x* and *y* components.

Stiffness

Stiffness is almost as important as strength. The stiffer a structure, the less it deflects under seismic force although, as noted previously, a smaller natural period of vibration caused by a stiffer structure will usually result in a structure attracting greater seismic force. Even though a building might be strong enough, if its stiffness is so low that it deflects excessively, its non-structural elements will still suffer damage (Chapters 10 and 11) and it will become prone to toppling. The more it deflects and its centre of gravity moves horizontally from its normal position, the more its own weight increases its instability (Fig. 2.14). For these reasons, design codes limit the maximum seismic deflections of buildings.

While the overall structural stiffness of a building is important, so is the relative stiffness of its different primary structural elements. In the example in Fig. 2.12, two identical structural elements resist seismic forces in each direction. Each wall resists half the total force. But what happens where the stiffness of vertical elements are different?

A key structural principle is that structural elements resist force in proportion to their stiffness. Where more than one member resists

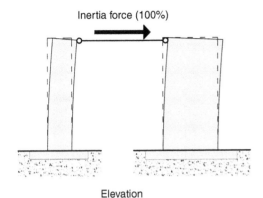

Inertia force (100%)

Elevation

1 m

1 m

2 m

Plan

11%   89%

Percentage of force resisted by each wall

▲ **2.15** Two walls of different stiffness and the force resisted by each.

forces the stiffer a member the more force it resists. Stiffness is proportional to the moment of inertia of a member (I). $I = bd^3/12$, b is the member width or breadth, and d its depth measured parallel to the direction of the force being resisted. Consider Fig. 2.15. Since both walls have the same width b, their respective stiffness is proportional to 1^3 and 2^3; that is, 1 and 8. The slender wall, therefore, resists 1/9th or 11 per cent of the force and the longer wall 8/9th or 89 per cent.

Where two such walls are the only force resisting structures in a certain direction, and they are located in plan along the same line, there is no structural problem. But if they are offset, as explained in the next section, the building experiences torsion. It twists in plan under seismic forces.

Torsion

Building torsion occurs either where structural elements are not positioned symmetrically in plan or where the centre of rigidity or resistance (CoR) does not coincide with the CoM.

Assume the building in Fig. 2.16(a) is single-storey with horizontal forces resisted by four identical square cantilever columns 1 m by 1 m, deliberately oversized to keep the arithmetic simple! Inertia forces acting uniformly over the whole of the roof plan are simplified as a single point force acting at the CoM, usually taken as the geometrical centre of the floor or roof plan. This force is resisted by the four columns. Because they are of identical stiffness each resists 25 per cent of the total force. The sum of all four column resisting forces acts along a line midway between the two column lines. The line of force through the CoM therefore coincides with the line of resistance through the CoR. The building is subsequently in both y direction and rotational equilibrium.

Figure 2.16(b) shows the right-hand columns now 2 m deep when considering their resistance in the y direction. The sum of the inertia force still acts at the CoM. (The influence of the increased weight of the larger columns moving the CoM to the right can be neglected because it is so small given the relatively heavy roof.) However, the CoR moves significantly to the right due to the increased stiffness of the right-hand side columns. From the considerations of the previous section the larger columns will resist 89 per cent of the force and the left-hand columns only 11 per cent. The position of the CoR

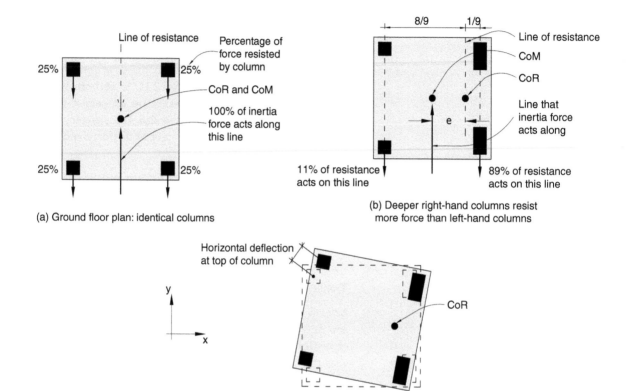

(a) Ground floor plan: identical columns

(b) Deeper right-hand columns resist more force than left-hand columns

(c) Twisting at roof level about the CoR

▲ **2.16** A symmetrical structure is modified to illustrate torsion and how it causes a building to twist. (Movement of the roof in the y direction is not shown.)

therefore lies at 1/9th of the distance between the two sets of column centrelines from the centreline of the right-hand columns. The lines of force and resistance are now offset by an eccentricity e (Figure 2.16(b)). This causes a torsion moment equal to the inertia force multiplied by e that twists the building clockwise in plan. Twisting occurs about the CoR (Fig. 2.16(c)). If the depths of the right-hand columns are further increased in the y direction, then the CoR moves further to the right, almost to the centreline of those columns, and increases the eccentricity to nearly half the building width.

The structural problems caused by torsion and the means of reducing them are discussed fully in Chapter 9. At this stage all that needs to be said is that torsion is to be avoided as much as possible. When a building twists, the columns furthest away from the CoR suffer serious damage due to excessive torsion-induced horizontal deflections.

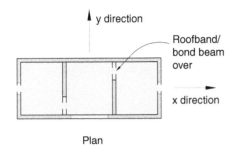

Plan

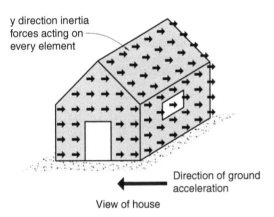

View of house

▲ **2.17** A simple building and y direction inertia forces.

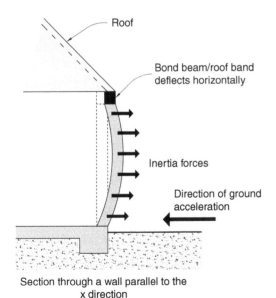

Section through a wall parallel to the
x direction

▲ **2.18** Effects of out-of-plane inertia forces on a wall.

FORCE PATHS

Architects and engineers determine force paths or load paths as they are also called by how they deploy structural elements and how those elements are joined and supported. The force path concept is a simple qualitative analytical tool for understanding and describing structural actions. Although it may not always give a complete picture of structural behaviour, it is useful in visualizing and comprehending structural behaviour, and is used extensively throughout this book.

A force path describes how forces within a structure are resisted by certain elements and transferred to others. The 'path' is the route we visualize forces taking as they travel from the applied forces to the foundations and into the ground beneath. The term 'force path' is metaphorical because forces don't actually move. Rather, they exist within structural members in a state of action and reaction in such a way that every structural element and connection remains in equilibrium.

Just because a force path can be described does not mean a structure is adequate. Every structural element and connection of a force path must be sufficiently strong and stiff to withstand the forces acting within them. Structural elements must fulfil two functions; first to resist forces, and second to transfer these forces to other members and eventually into the ground. The adequacy of a force path is verified by following it step-by-step, element-by-element. Three questions are addressed and answered at each step – what resists the force and how, and where is it transferred to?

Consider the force paths of a simple single-storey building with two interior walls (Fig. 2.17). Earthquake accelerations in the y direction induce inertia forces in all building elements namely the roof and walls that need to be transferred to the ground.

Walls parallel to the x direction require sufficient bending and shear strength to function as shallow but wide vertical beams. They transfer half of their own inertia forces up to the roof band and the other half down to the foundations (Fig. 2.18).

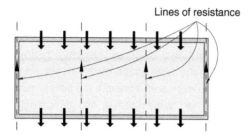

Lines of resistance

Eaves level wall and roof forces acting
on the bond beam or roof band

Wall band beam or roof band

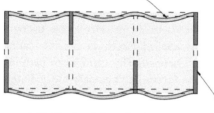

Bond beams deflect horizontally under the
action of the inertial forces from the roof and
the walls and transfer them to the shear walls
orientated parallel to the direction of ground
shaking.

y direction
shear wall

Resisting shear walls

▲ **2.19**   Bond beams or roof bands distribute inertia forces at
eaves level to shear walls parallel to the y direction.

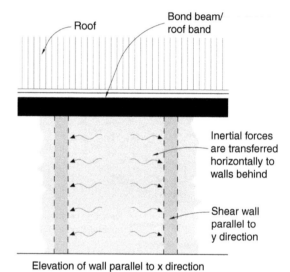

Roof

Bond beam/
roof band

Inertial forces
are transferred
horizontally to
walls behind

Shear wall
parallel to
y direction

Elevation of wall parallel to x direction

▲ **2.20**   Force paths for a short length of out-of-plane loaded
wall restrained by walls at right-angles.

Roof forces are resisted and transferred by roof structure down to
the bond beam at eaves level. In the absence of a ceiling diaphragm
which could also transfer the roof forces horizontally, the bond beam
resists and transfers roof and wall inertia forces to the shear walls acting in the y direction. The bond beam deflects horizontally, functioning
as a continuous horizontal beam (Fig. 2.19). The walls parallel to the x
direction have little or no strength against y direction forces or out-of-
plane forces except to span vertically between foundations and bond
beam. They are usually not strong enough to cantilever vertically from
their bases so they need support from the bond beam.

If the walls parallel to the y direction were more closely spaced in plan,
say 2 m or less apart, then out-of-plane forces acting on the walls at
right angles if of concrete or masonry construction can take a short-
cut. They travel sideways, directly into those walls parallel to the y
direction (Fig. 2.20). In this case the bond beam resists little force from

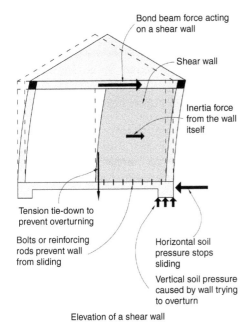

Bond beam force acting on a shear wall

Shear wall

Inertia force from the wall itself

Tension tie-down to prevent overturning

Bolts or reinforcing rods prevent wall from sliding

Horizontal soil pressure stops sliding

Vertical soil pressure caused by wall trying to overturn

Elevation of a shear wall

▲ **2.21** Inertia forces and resisting actions on a shear wall.

Roof plane bracing from apex to bond beam

Inertia forces act on the gable and roof

Roof Plan

y

x

Bond beams deflect as they transfer x direction inertia forces to shear walls

Shear wall in x direction

Plan at bond beam level

▲ **2.22** Bond beams distribute x direction inertia forces at roof level to shear walls.

the short out-of-plane laden wall. If walls are of light-timber frame construction no matter how closely spaced the cross-walls are wall studs will always span vertically and half of the wall inertia force will be transferred upwards to the bond beam.

At this stage of the force path, *y* direction inertia forces that arise from the roof and walls running in the *x* direction are transferred by the bond beams to the four lines of shear walls. When a wall resists a force parallel to its length, that is an in-plane force, it functions as a shear wall. Bond beams over the *y* direction walls acting in either tension or compression transfer forces from the roof and walls into these *y* direction walls. Due to their strength in bending and shear they then transfer those forces from the bond beams, plus their own inertia forces, down to the foundations (Fig. 2.21). Overturning or toppling of walls is prevented by a combination of their own weight and connection to walls at right angles as well as by ties or bolts extending into the foundations.

Finally, consider shaking in the *x* direction. In a real earthquake this happens simultaneously with *y* direction loading. Similar force paths apply except for two differences. First, the gable ends, which are particularly vulnerable against out-of-plane forces due to their height, need to be tied back to roof structure. Second, bracing is required in the roof plane to resist inertia forces from the top of the gables as well as the inertia force from the roof itself (Fig. 2.22). The bracing transfers these forces through tension and compression stress into the *x* direction bond beams. From there, forces travel through the four *x* direction shear walls down to the foundations.

The force paths for shaking in both orthogonal directions have been described. During a quake with its directionally random and cyclic pattern of shaking both force paths are activated at the same time. This means that many elements simultaneously resist and transfer two different types of force. For example, walls resist out-of-plane forces while also acting as shear walls. Earthquake shaking induces a very complex three-dimensional set of inertia forces into a building but provided adequate force paths are provided as discussed, building occupants will be safe and damage minimized.

NOTES

1 A bending moment acting upon a member causes it to bend or flex. It is calculated at a section through a member by summing all the moments (forces times their distance) acting to one side of the section. As bending occurs, internal tension and compression bending stresses develop within the member. A bending moment diagram is a graph that shows how the intensity of bending moment varies along a member.

2 A shear force at a section through a member is the sum of forces acting at right angles to its length to one side of that section. Shear forces induce a shearing action or shear stresses normal to the member length. A shear force diagram is a graph that shows how the intensity of shear force varies along a member.

3 SEISMIC DESIGN APPROACHES

INTRODUCTION

Having considered the basic principles of seismic resistance in Chapter 2, we now step back and take a wider perspective to examine the current philosophy of seismic design. This chapter begins with a brief historical overview of earthquake resistant design, outlining some of the key developments directly relevant to the seismic design of buildings. This is followed by a review of the philosophy of seismic design as generally adopted internationally. Several important architectural implications are briefly noted before a concluding discussion on ductility.

Ductility, one of the principal concepts of contemporary seismic resistant design practice, was introduced in Chapter 2. Its application for each of the main structural seismic resisting systems – structural walls, cross-braced frames and moment frames – is described in Chapter 4. However, before that detailed examination of ductility this chapter explains in general terms how architects and structural engineers achieve ductile structures; that is, structures that endure earthquake shaking strong enough to exceed their strength without collapsing. A final example illustrates the steps involved in designing a simple ductile reinforced concrete structure.

HISTORICAL OVERVIEW

Claims of earthquake awareness informing the design and construction of vernacular architecture in long-established communities are not uncommon. But it is unusual for a specific example from such a tradition to stimulate contemporary seismic innovation. A reviewer of the Renzo Piano Building Workshop's new Hermès building in the seismically-prone city of Tokyo explains: 'At 50 m tall and with a main structural span of only 3.8 m, the unusual slenderness of the structure results in high overturning moments during an earthquake and high levels of tension in the columns. The engineer, Ove Arup & Partners, found inspiration in the tall, thin wooden Buddhist pagodas of Japan.

Records show that in the past 1400 years, only two have collapsed – believed to be because the columns are discontinuous from floor to floor'.[1] In the Hermès building a similar principle was adopted by allowing columns to lift off the foundations during a quake and activate energy-absorbing dampers in the process.

In spite of examples of admirable seismic performance being cited from antiquity, especially where wood construction was employed, the history of earthquake resistant design based upon modern scientific methods is extremely short. While the wonders of Egyptian architecture date from 3000 BC, five thousand years ago, the beginnings of modern earthquake engineering practice emerged from the earthquake devastation of two cities at the turn of the twentieth century, only one hundred years ago. The 1906 San Francisco, California earthquake and the 1908 earthquake in Reggio and Messina, Italy in particular were pivotal in stimulating scientific enquiries that have led to what has now become modern earthquake engineering.[2]

▲ **3.1** Damage to the city of Messina from the 1908 earthquake.
(Reproduced with permission from Arturo Tocchetti; copyright Russell Kightley).

Some earthquake resistant features had been introduced into buildings prior to the 1906 San Francisco event,[3] but the main advances immediately following it were in the field of seismology including the development of instruments to record ground shaking. The 100,000 plus fatalities in southern Italy (Fig. 3.1) and the fact that some buildings survived the quake spurred on engineers who for the first time developed a method of designing buildings based upon Newton's Second Law of Motion described in Chapter 2.

After the 1923 Kanto earthquake in Japan and its estimated 140,000 fatalities most of whom were killed by post-earthquake fires, some successful applications of earthquake engineering focused the international spotlight on Japanese developments. These influenced researchers and designers in the USA whose efforts were further stimulated by the 1925 Santa Barbara, California earthquake. The iterative yet progressive nature of earthquake design developments continued. Field measurement of earthquake shaking and damage observation and its analysis was followed by laboratory testing. Understanding subsequently deepened by mathematical (now computer) modelling led to code modifications which were reviewed in the light of more recent earthquake data, and so the step-by-step refinement of seismic design practice continued.

Bertero and Bozorgnia note that as a consequence of the Santa Barbara event 'with the cooperation of many engineers and architects, the Pacific Coast Building Officials Conference adopted the Uniform Building Code (UBC)'.[4] This 1927 Code included the first seismic design guidelines in the USA, but the first *enforced* seismic code was the 1933 Los Angeles City Code. Some ten years later its provisions acknowledged the importance of building flexibility, but it was not until 1952 that the natural period of vibration, the primary dynamic characteristic of a building that is taken for granted today, was explicitly incorporated into a code.

Up until the mid-1950s, structural engineers focused their attention upon providing buildings with sufficient strength and stiffness to meet code defined force levels. Their understanding and application of relatively new earthquake engineering technology had yet to grapple with what is now considered to be a crucial issue; namely, what happens to a structure if its inertia forces exceed those for which it has been designed? The concept of ductility was first codified in a general form by the Structural Engineers Association of California in 1959, but it wasn't until approximately ten years later that several New Zealand structural engineers proposed a method whereby ductile structures of any material could be reliably designed and built.[5]

This method, known as Capacity Design is a design approach that is explained fully in the following section. Briefly, it imposes a hierarchy of damage upon a structure so even when inertia forces exceed design values, damage is concentrated in less vital sacrificial members. Other members more critical to the survival of a building – like columns – suffer little or no damage. Once columns are damaged they may not be able to support the weight of construction above and, as so often occurs during damaging quakes, buildings collapse.

Capacity Design to a greater or lesser extent is now well established in the world's leading seismic design codes. It is an important component of state-of-the-art seismic design practice. Although rigorously developed through extensive laboratory tests and computer simulations, it is sobering to reflect that at the time of writing Capacity Designed buildings have yet to be put to the ultimate test – an earthquake at least as strong if not stronger than a design-level earthquake. While Capacity Design offers a design methodology capable of preventing building collapse, an awareness of the need to protect the fabric and contents of buildings from earthquake damage is growing. Whereas former codes concentrated on saving lives, now a more holistic appreciation of the seismic performance of buildings – including reducing non-structural damage and post-earthquake disruption – is gaining greater emphasis in codes.

Unfortunately, Capacity Design was not incorporated into building designs until the early 1970s in New Zealand and later in other countries. Therefore, the vast majority of buildings in all cities pre-date the current codes that attempt to achieve buildings with dependable structural ductility. This large proportion of the building stock is therefore vulnerable to serious brittle damage, as observed in the 1994 Northridge, California and the 1995 Kobe, Japan earthquakes and more recent earthquakes elsewhere (Figs. 3.2 and 3.3). This explains why the subject of retrofitting or improving the earthquake performance of existing buildings, as discussed in Chapter 12, is so relevant.

Some readers may have noticed that there is almost no mention of architects cited throughout the short history above. Did architects contribute to those developments? Apparently not, according to the histories that focused upon the seismological and engineering advances of those times. Certainly, architects *were* involved to some degree; for example, members of code review committees. But it was probably the 1971 San Fernando, California earthquake that highlighted, albeit rather negatively, the importance of the architect in achieving good seismic performance. In the quake's aftermath the infamous failures at the newly commissioned Olive View Hospital attracted widespread attention (Fig. 3.4). Strong non-structural elements in the form of masonry infill walls precipitated the collapse of several elevator towers and were responsible for serious structural damage to the main block, necessitating its demolition. Now the

▲ 3.2 Partial collapse of a car parking garage, Los Angeles. 1994 Northridge, California earthquake.
(Reproduced with permission from A.B. King)

▲ 3.3 Brittle structural damage, Kobe. 1996 Kobe, Japan earthquake.
(Reproduced with permission from Adam Crewe).

▲ **3.4** Damage to the Olive View hospital. 1971 San Fernando, California earthquake. (Reproduced with permission from Bertero, V. V. Courtesy of the National Information Service for Earthquake Engineering, EERC, University of California, Berkeley).

architect's role in contributing to sound seismic building configuration is widely recognized. This crucial aspect of seismic resistant design receives detailed treatment by Christopher Arnold and Robert Reitherman in their classic book,[6] in other publications authored by Christopher Arnold and others, and is a re-occurring theme of this book.

But let us return to the previous question regarding architects' roles in advancing the practice of earthquake-resistant design. It appears that their input is not to be found in contributions to particular technical developments but rather in their eagerness to adopt new structural forms, especially moment frames. Architects from the Chicago School at the end of the nineteenth century were quick to escape the architectural restrictions imposed by load-bearing masonry walls. They embraced iron and then steel rigid framing. Frames not only offered greater planning freedom epitomized by the 'free-plan' concept, but also extensive fenestration and its accompanying ingress of natural light. Although rigid frames have functioned as primary structural systems in buildings for over a hundred years, as mentioned previously, only those built after the mid-1970s comply with strict ductility provisions and, therefore, can be expected to survive strong shaking without severe structural damage.

Structural engineers sometimes struggle to keep pace with architects' structural expectations. Consider Le Corbusier's influential 1915 sketch of the Dom-Ino House, a model of simplicity and openness (Fig. 3.5). While structurally adequate in seismically benign

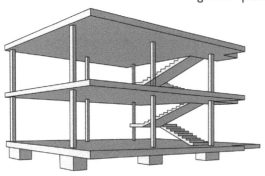

▲ **3.5** Reinforced concrete structure of the Dom-Ino House, Le Corbusier, 1915.

▲ **3.6** Villa Savoye, Poissy, Paris, Le Corbusier, 1929. This villa incorporates all of Le Corbusier's 'five points' of architecture. A combination of piloti and irregularly placed concrete block walls resist horizontal forces.

environments, its structural configuration is inappropriate for seismic resistance. Its columns are too weak to cantilever two storeys high from the foundations, an absence of beams prevents reliable moment frame action, and the reinforced concrete stair induces in-plan torsion. Further seismic weaknesses are introduced by masonry infill walls. Sadly, these inherent seismic deficiencies have been unwittingly ignored by structural engineers who have assisted architects to achieve Le Corbusier-inspired design concepts in seismically active regions. Reflecting upon the destruction of the El Asnam modern concrete buildings during the 1980 Algerian earthquake, architect Marcy Li Wang points out that every one of Le Corbusier's 'five points of a new architecture' that have been widely embraced by architects worldwide, leads to seismic deficiencies (Fig. 3.6).[7] 'While "the five points" have set generations of architectural hearts beating faster, they have more sinister overtones for structural engineers and other seismic specialists who would recognize pilotis as 'soft stories' that have been the failing point of dozens of modern buildings in earthquakes all over the world.'[8] Architects need to understand how building configuration affects seismic performance. It is unrealistic to expect that engineers can somehow design poorly architecturally configured buildings to perform well in moderate to severe earthquakes.

CURRENT SEISMIC DESIGN PHILOSOPHY

One of the themes emerging from a history of seismic resistant design is that of international collaboration. The following review of the current philosophy of seismic design acknowledges that theme as it draws upon the earthquake provisions of each of the four countries or regions that at various times have provided leadership in the development of modern codes. Relevant aspects of codes from Japan,[9] Europe,[10] USA[11] and New Zealand[12] are referenced to offer an international perspective.

Although there are many points of detail on which the codes differ, taken together they present a reasonably united philosophy to outwit quakes. Readers are encouraged to check how the following points align with those of their own earthquake code. Since structural engineers are the intended readership of codes, these codes are not particularly accessible to other professionals. Hence, readers will appreciate structural engineering assistance with code way-finding, interpretation and the answering of questions related to their local situations.

One general comment at this point about all codes is that they proscribe *minimum* standards. As discussed later, particularly in Chapter 13, there are some projects where architects recommend to clients that higher than the minimum standards of seismic resistance be adopted.

Design-level earthquakes

Seismic resistant design is intended to achieve two objectives:

- Protect human lives, and
- Limit building damage.

The first objective is achieved primarily by the provision of adequate strength and ductility. This ensures that a building is protected from full or partial collapse during large earthquakes that occur infrequently. The second objective limits building damage during lesser, more frequently occurring earthquakes, in order to minimize economic losses including loss of building functionality.

A code design-level earthquake is defined as one with an average reoccurrence interval of approximately 500 years. A building with a design life of 50 years has therefore approximately a 10 per cent chance of experiencing a design-level earthquake with that return period (refer to Poisson's equation in Chapter 1). Codes specify the intensity of design accelerations appropriate for that magnitude of earthquake through their response spectra (see Fig. 2.9(b)). Collapse is to be avoided during the design-level earthquake but considerable structural and non-structural damage is usually considered acceptable. Those lengths of structural members that have functioned as ductile structural fuses may be badly damaged but due to careful detailing they are expected to maintain most of their original strength and be repairable. The maximum permissible horizontal deflections of buildings during the duration of strong shaking are limited by codes to control damage and prevent overall instability. But there is still a risk that an especially strong pulse might cause permanent deformations that are so large in some buildings as to require their demolition.

The situation described above can be restated as follows: over a fifty-year period that could approximate its design life, a building has a 10 per cent chance of experiencing the design-level or a larger earthquake. The intensity of shaking depends upon both the regional seismic zone in which the building is located and the underlying ground conditions. During the design-level earthquake the structure and entire fabric of the building, including its contents, will almost certainly be seriously damaged. Lives will not be lost but post-earthquake entry may be prohibited by civil defence personnel and the building may require demolition. This is the rather depressing reality of the scenario where a building designed to code requirements has 'survived' the design-level earthquake. Current design approaches aim to prevent collapse but not damage during such a large event. No wonder some clients request enhanced performance (Chapter 13) and researchers are busy investigating ductile but damage-free structural systems (Chapter 14).

Although structural and building fabric damage is not required to be prevented during a large earthquake, designers are obliged to avoid damage during small earthquakes that occur relatively frequently (Fig. 3.7). A second and smaller design earthquake with a return period in the range of 25 to 50 years is the maximum event for which damage is to be avoided. These short return period events represent an 86 per cent and 63 per cent likelihood of occurrence, respectively, during an assumed fifty-year design life of a building. Since building damage correlates strongly with the amount of horizontal deflection in any storey, or interstorey drift, codes limit the maximum deflections during frequently occurring earthquakes. Apart from unrestrained building contents that may be damaged no structural or non-structural building elements are expected to require repair.

▲ **3.7** Plywood sheets cover broken windows after the small 2001 Nisqually Earthquake, near Seattle.
(Reproduced with permission from Graeme Beattie).

The previous two paragraphs apply to typical buildings in a community, like those accommodating apartments, shops or offices. But what about especially important buildings like schools, hospitals, and fire stations? Since societies expect them to perform better, the strengths of these facilities are increased beyond that of less critical buildings by factors of up to 1.8. This strength enhancement enables them to survive the code design-level earthquake with significantly less damage than that incurred by a typical building. Damage limitation requirements for these important or critical facilities are also more rigorous. Full building functionality is expected immediately following an earthquake with a return period far in excess of 50 years.

Ductility

During the initial seconds of a design-level earthquake, when a structure vibrates elastically with normal amounts of damping, the consequent inertia forces become large. As mentioned in Chapter 2, because earthquake loading is cyclic and a structure usually possesses some ductility, structural engineers reduce the design-level seismic forces to well below those that would occur if the structure were to continue to remain elastic. Ductility is a measure of how far a structure can safely displace horizontally after its first element has been overstressed to the extent its steel yields or fibres, in the case of wood structures, begin to rupture (Fig. 3.8). The degree of ductility indicates the extent to which earthquake energy is absorbed by the structure that would otherwise cause it to continue to resonate. Those areas of structures designed to absorb or dissipate energy by steel yielding are called *structural fuses* or *plastic hinges*.

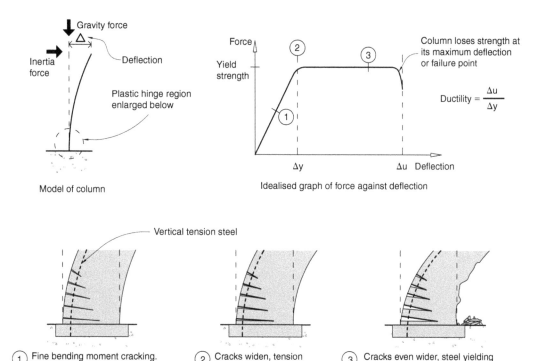

Model of column

Idealised graph of force against deflection

$$\text{Ductility} = \frac{\Delta u}{\Delta y}$$

Vertical tension steel

1. Fine bending moment cracking. Vertical tension reinforcement is still in the elastic range.

2. Cracks widen, tension steel starts to yield.

3. Cracks even wider, steel yielding and cover concrete spalled off. Typical damage in a plastic hinge.

▲ **3.8** A vertical reinforced concrete cantilever column subject to a horizontal force. The force-deflection graph and the damage states reflect the ductility of the structural fuse or plastic hinge region.

Codes allow designers to reduce the inertia forces likely to occur in a design-level earthquake in proportion to the ductility a given structural system might possess. A huge reduction in design force is allowed for high-ductility structures. Due to such low design forces that are as little as one-sixth of the design force for a brittle structure and the consequent low structural member strengths required, relatively shallow beams and slender columns are achievable. But the disadvantages of a high-ductility design include the creation of a more flexible structure that will sustain more structural and non-structural damage than a stronger and stiffer alternative. Smaller members usually result in lower construction costs. But those savings are somewhat offset by the special detailing required in the structural fuse regions of ductile structures to allow yielding and plastic action without excessive damage or loss of strength.

Alternatively, designers can opt for a low-ductility structure. Due to its lesser ductility its design forces are higher. Consequently, structural members have to be stronger and larger, with obvious architectural implications. Its increased structural footprint, with perhaps deeper columns or longer walls, may now not integrate well with the desired internal layout.[13] While structural dimensions are larger than those of a high ductility structure, the advantages to a client are a stiffer and stronger building with less damage expected to non-structural and structural elements. Also less sophisticated and costly structural detailing is necessary at locations likely to suffer damage. They are expected to be far less severely damaged in the design-level earthquake compared with similar regions within a high-ductility structure.

Some codes permit a structure to remain elastic through the design-level earthquake. This is a sound option if, other than for structural reasons, a larger than the required amount of structure is readily available to resist seismic forces. This strategy is often employed where two boundary walls provide fire resistance (Fig. 3.9). The walls are probably far stronger than needed for forces acting parallel to their lengths and, therefore, can resist elastic response force levels that are not reduced by ductility considerations. In this situation, structural detailing can be kept at its most simple and cost-effective, although structural engineers must consider the consequences of an earthquake whose intensity exceeds that of the design-level earthquake.

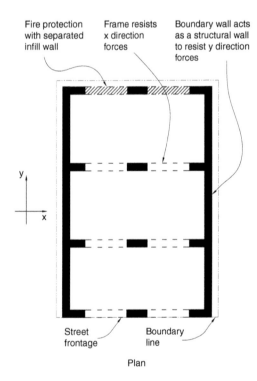

Fire protection with separated infill wall

Frame resists x direction forces

Boundary wall acts as a structural wall to resist y direction forces

y

x

Street frontage

Boundary line

Plan

▲ **3.9** Plan of a building whose structural walls on the boundary provide excess seismic strength in the y direction. Frames resist x direction forces.

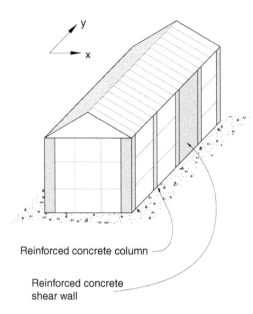

Reinforced concrete column

Reinforced concrete shear wall

▲ **3.10** Single-storey building with x direction horizontal forces resisted by cantilever columns. Shear walls resist y direction forces.

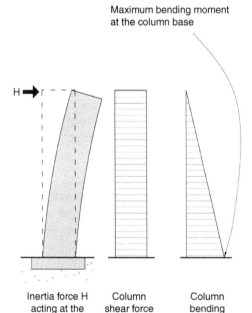

Maximum bending moment at the column base

H

| Inertia force H acting at the top of a column | Column shear force diagram | Column bending moment diagram |

▲ **3.11** The inertia force deflects the column and produces column shear force and bending moment diagrams (The bending moment is drawn on the compression-side of the column rather than the tension-side as is customary in some countries).

Capacity Design

It is relatively easy for structural engineers to increase the structural strength of a member, or even that of a complete structural system. Additional reinforcing bars can usually be added to a reinforced concrete member. In steel or timber construction, substitution of a larger cross-sectional area increases member strength. Unfortunately, it is nowhere near as easy to increase ductility even in a steel structure. So, how are ductile structures designed? How can structural collapse be prevented if earthquake shaking exceeds the strength of the design-level earthquake?

A ductile structure is designed using the Capacity Design approach. This involves the following three steps:

- Choose how the structure is to deflect in a seismic overload situation so that the structure is able to absorb sufficient earthquake energy before it deflects to its limit.
- Provide a hierarchy of strength between and within structural members to allow structural fuses or plastic action only in non-critical members and to prevent brittle failure occurring anywhere, and finally,
- Detail structural areas that are intended to act as fuses so they avoid severe damage and excessive loss of stiffness and strength.

To understand the application of Capacity Design, consider a simple single-storey building (Fig. 3.10). Assume simply-supported steel or timber trusses span between the pairs of reinforced concrete columns. The *x* direction seismic forces on the building are resisted by columns functioning as vertical cantilevers. Two concrete shear walls, not considered further in this example, resist *y* direction seismic forces. The seismic force acting on a column and its internal structural actions such as bending moments and shear forces are shown in Fig. 3.11.

Before applying the Capacity Design approach, we pause to explore the following question: If the earthquake force exceeds the strength of the column how will the column be damaged? The correct answer depends on how the column is reinforced and the strength of its foundations (Fig. 3.12).

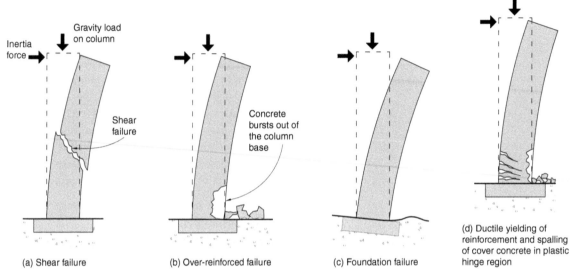

(a) Shear failure

(b) Over-reinforced failure

(c) Foundation failure

(d) Ductile yielding of reinforcement and spalling of cover concrete in plastic hinge region

▲ **3.12**   Four potential types of failure for a cantilever column. The only ductile and desirable overload mechanism (d) occurs if the other three are suppressed by making them stronger than the force to cause (d).

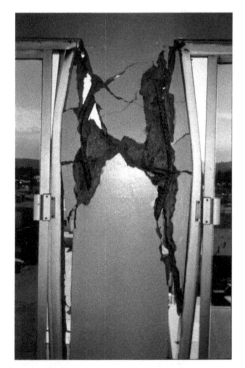

▲ **3.13**   Brittle reinforced concrete column shear failure. Los Angeles, 1994 Northridge earthquake. (Reproduced with permission from A.B. King).

The first and most likely possibility is that the column will suffer shear failure (Fig. 3.12(a)). If the shear strength of the column is less than the strength at which any of the other damage modes occur, a sudden brittle diagonal shear crack forms. The column almost snaps in two. The crack creates an inclined sliding plane which greatly reduces the ability of the column to support vertical loads. Shear failures are often observed in quake-damaged columns not designed in accordance with the Capacity Design approach (Fig. 3.13).

A second and less likely scenario is where excessive vertical reinforcing steel is placed in the column. It is 'over reinforced'. Provided no other type of damage occurs beforehand, as the bending moment at the base of the column increases due to increasing seismic force and thus stretches the left-hand side vertical reinforcing steel in tension, compression stress builds up within the concrete on the other side. Because of the large amount of vertical steel acting in tension the concrete under compression finds itself the weaker element. As often demonstrated to classes of civil engineering students the over-stressed concrete suddenly and explosively bursts and the column falls over (Fig. 3.12(b)).

The foundation system may also sustain possible damage or collapse. Either the foundation soil might be too weak to support the combined vertical and horizontal stress under the column footing or the footing might overturn because it is undersized (Fig. 3.12(c)). Both of these mechanisms lead to severe building damage.

The final type of damage is shown in Fig. 3.12(d). Once again, assume no other prior damage has occurred to the column. As the inertia force at its top increases, the bending moment at the column base causes cracks in the concrete and increases the tension stress in the vertical reinforcing steel until it begins to yield, or in other words enters the plastic range. The maximum horizontal force the column can sustain has been reached. No additional force can be resisted so if the inertia force continues to act on the column, cracks grow wider as the reinforcing steel yields plastically. The area at the base of the column with its wide cracks and where the steel has been strained plastically is a structural fuse region, often called a *plastic hinge zone*. Even though the vertical steel has stretched plastically in tension it is still strong. Due to the high compression stress from the bending moment, as well as the gravity load on the column, the cover concrete spalls from the compression side. This damage is not serious. The column is only slightly weaker and damage to cover concrete can be repaired quite easily. *Of the four types of damage this is the only one that can be described as ductile.*

We now return to the three steps of Capacity Design. Since we have identified a ductile overload damage mechanism, namely ductile bending deformation at the column base, we select it as the desired mode of damage. Then by applying the remaining two design steps described below, we ensure that that type of damage and that alone occurs in the design-level event. Tensile yielding of the column vertical steel absorbing earthquake energy is the source of ductility.

The next Capacity Design step involves providing a hierarchy of strength. All undesirable types of damage such as shear and foundation failure must be prevented. So, disregarding the value of design force acting on the column, we calculate the bending strength of the column using the actual area of vertical steel provided. This bending strength is usually greater than the design bending moment because of the need to round-up the number of reinforcing bars to a whole number during the design process. Then the actual bending strength is increased by a small factor of safety to determine the maximum possible bending strength. This acknowledges that reinforcing steel and concrete are usually stronger than their minimum specified strengths.

Once the maximum possible bending strength at the column base is calculated, the damage modes of Fig. 3.12(a) and (c) can be prevented by ensuring they occur at a higher level of seismic force. Column ties of sufficient diameter, and close enough vertical spacing between ties, completely prevent shear failure prior to the maximum possible bending strength occurring. The column footing is also dimensioned using the maximum possible bending strength of the column. Overturning and foundation soil failure is therefore prevented. Finally, the structural engineer checks if the column section is 'over reinforced'. If so, the column depth is increased and the amount of reinforcing steel reduced.

The final step of the design process involves detailing the structural area designated as the fuse region. In this example it means (1) adding extra horizontal reinforcing ties to stop the main vertical bars from buckling in compression after the cover concrete has spalled and (2) decreasing the vertical spacing of the ties so as to confine the concrete in the fuse or plastic hinge region more firmly. It is like applying bandages to stabilize the reinforcing bars and confine the cracked concrete so it doesn't fall out from the column core (Fig. 3.14). Figure 3.15 shows an example of a plastic hinge in a column where spiral reinforcement has been effective in confining the concrete.

Once all of the Capacity Design steps are completed construction commences and the structure is ready to be put to the test. If, for architectural or other reasons, high ductility was assumed in order to achieve slender columns, inertia forces from a moderate to severe earthquake will definitely exceed the design bending strength of the columns. Plastic hinges will form, accompanied by concrete cracking and spalling of cover concrete. Because all undesirable brittle types of failure are prevented by virtue of being stronger than the ductile mechanism, ductile behaviour is ensured. The column definitely suffers damage but is repairable. Any loose concrete is removed and fresh concrete cast. Careful inspection will reveal whether or not cracks need epoxy grouting. Fallen cover concrete is reinstated with cement plaster. The column is ready for the next quake!

The preferred ductile mechanisms and the architectural implications of structural systems other than the vertical cantilever columns considered above such as structural walls, cross-braced frames and moment frames are discussed in the following chapter.

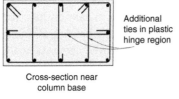

Additional ties in plastic hinge region

Cross-section near column base

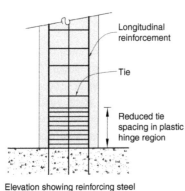

Longitudinal reinforcement

Tie

Reduced tie spacing in plastic hinge region

Elevation showing reinforcing steel

▲ **3.14** Additional reinforcing steel to confine the concrete in the fuse or plastic hinge region of a column.

▲ **3.15** A column plastic hinge. Closely-spaced spiral ties confine the column core.
Cover concrete has spalled off. Mexico City, 1985 Mexico earthquake.
(Reproduced with permission from R.B. Shephard).

REFERENCES AND NOTES

1 Heneghan, T. (2001). Japanese lantern. *Architectural Review*, **201**:1255, 78–81.

2 Reitherman, R. (2006). Earthquakes that have initiated the development of earthquake engineering. *Bulletin of the New Zealand Society for Earthquake Engineering*, **38**:3, 145–157.

3 Tobriner, S. (2006). *Bracing for disaster: earthquake-resistant architecture and engineering in San Francisco, 1838–1933, p. 86.* Heyday Books, Berkeley, California, p. 86.

4 Bertero, V.V. and Bozorgnia, Y. (2004). The early years of earthquake engineering and its modern goal. In *Earthquake Engineering: From engineering seismology to performance-based engineering*, Bertero, V.V. and Bozorgnia, Y. (eds). CRC Press, pp. 1-1–1-17.

5 Reitherman, R. (2006). *Connections: The EERI oral history series: Robert Park–Thomas Paulay*. Earthquake Engineering Research Institute, Oakland, California, p. 33.

6 Arnold, C. and Reitherman, R. (1982). *Building Configuration and Seismic Design*. John Wiley & Sons, Inc.

7 Le Corbusier's five points of architecture are: 1, Pilotis or free-standing ground floor columns to free-up the ground floor; 2, A free plan (no load-bearing walls); 3, A free façade (non-structural external skin); 4, Roof terraces; 5, Strip windows.

8 Li Wang, M. (1981). Stylistic dogma vs. seismic resistance: the contribution of modernist tenets to an Algerian disaster. *AIA Journal*, 59–63.

9 Midorikawa, M. et al. (2003). Performance-based seismic design code for buildings in Japan. *Earthquake Engineering and Engineering Seismology*, **4**:1, 15–25.

10 British Standards (2004). *Eurocode 8: Design of Structures for Earthquake Resistance – Part 1: General Rules, Seismic Actions and Rules for Buildings*, BS EN 1998-1:2004. British Standards.

11 ASCE (2005). *Minimum Design Loads for Buildings and Other Structures*, ASCE/SEI 7-05. American Society of Civil Engineers.

12 Standards New Zealand (2004). *Structural Design Actions, Part 5: Earthquake Actions – New Zealand*, NZS 1170.5:2004. Standards New Zealand.

13 The term 'structural footprint' refers to the cross-sectional area of structure at ground floor level where structural dimensions of columns and walls are largest.

4 HORIZONTAL STRUCTURE

INTRODUCTION

Some readers may be surprised to find a chapter devoted to the horizontal structure necessary for seismic resistance. After all, images of earthquake damaged buildings invariably feature damaged walls and columns. Also, in the design process an architect's awareness of vertical elements, like columns and shear walls, is heightened due to their influence on architectural requirements such as circulation, provision of natural light and spatial functioning. These structural elements make their presence felt in both plan and section – but what about horizontal structure? What is this structure, and where is it shown in architectural drawings?

Just because most horizontal seismic resisting structure is unseen or not perceived as such, and because it multi-tasks with other structural elements like floor slabs and ceilings, it doesn't mean it is unimportant. In fact it is essential. It is a vital component of every seismic force path. Although it may not grace architectural plans, the structural requirements of horizontal structure may have affected significant architectural decisions, such as the number of shear walls and the spacing between them. As for its apparent lack of damage during earthquakes, that is due mainly to its often inherent strength. Earthquakes damage weaker vertical structural elements first. Ironically, horizontal structure is quite unnecessarily sacrificially protected by damage to vertical structure which can, of course, lead to building collapse.

The discussion of force paths in Chapter 2 explains how inertia forces first move horizontally before being channelled vertically downwards towards the foundations. Horizontal structure, an essential participant in the sequence of force transfer within a building is, therefore, considered as a primary structure in any seismic force path. For that reason, this explanation about horizontal structure precedes the following chapter on vertical structure (which includes consideration of shear walls, cross-braced frames and moment frames). While diaphragms constitute the predominant horizontal seismic force resisting structural element, the structural necessity of collectors, ties and bond-beams and their possibly significant architectural implications are also discussed.

DIAPHRAGMS

Consider the floor plan in Fig. 4.1. Imagine it to be a typical floor of a medium-rise building. For most of its design life the floor structure

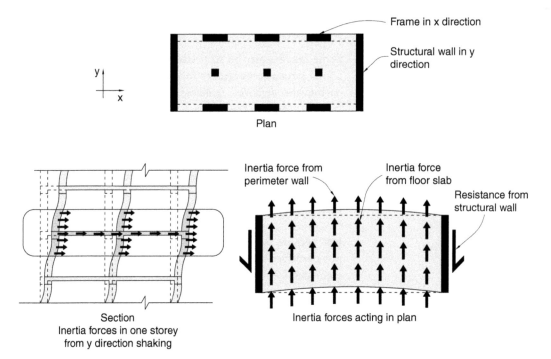

▲ **4.1**   Inertia forces within a multi-storey building shown in plan and section.

resists gravity forces; dead and imposed forces that act vertically. But during an earthquake, that perhaps lasts only between 10 to 100 seconds, the floor structure resists horizontal seismic forces. During this relative infinitesimally brief period of time, when the floor structure is called upon to resist not only gravity but also horizontal forces, it is described as a *diaphragm*. When the ground shakes in the y direction, inertia forces are induced in exterior and interior walls and the floor slab itself. Usually inertia forces acting upon a wall that is loaded

▲ 4.2 The weakness of unreinforced masonry walls and their connections prevented inertia forces being transferred safely to diaphragms at first floor and roof level. Santa Monica, 1994 Northridge earthquake.
(Reproduced with permission from A.B. King).

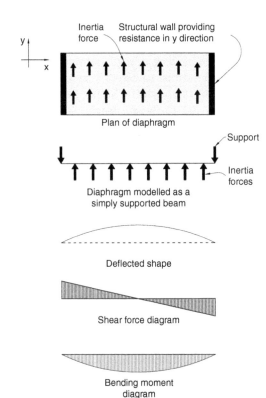

▲ 4.3 A diaphragm as a simply-supported beam.

perpendicular to its length that are called *face-loads* or *out-of-plane forces* are transferred vertically, half up and half down to diaphragms, which then transfer them to vertical structure acting in the direction of shaking; in this case two structural walls. The walls then transfer the inertia forces to the foundations. Diaphragms play an identical role when wind forces act on a building.

Strong and ductile connections between walls and diaphragms are necessary. Where connections are lacking, or are brittle or weak as in many existing unreinforced masonry buildings, walls fall outwards from buildings (Fig. 4.2). For example, during the Northridge earthquake, inadequate connection details between walls and roof diaphragms of newer buildings also led to walls collapsing.

When functioning as a diaphragm, a floor slab acts like a beam albeit resisting *horizontal* rather than *vertical* forces and possessing a span-to-depth ratio much smaller than that of a typical beam. Just like a simply-supported beam the diaphragm bends under the influence of the horizontal inertia forces, spanning not between piers or posts, but in this case between two structural walls. It experiences bending moments and shear forces whose distributions along its length are identical to that of a gravity-laden beam. So a diaphragm is modelled just the same except that we need to remember that the direction of force and bending is horizontal (Fig. 4.3).

A diaphragm is therefore a beam that acts horizontally. As such, it requires stiffness and strength. Its often squat geometry avoids excessive horizontal deflections. However, a structural engineer needs to check that a diaphragm slender in plan is not too flexible, particularly if called upon to resist and transfer inertia forces from heavy masonry walls orientated parallel to its length (see Fig. 6.10).

The maximum horizontal deflection of a typical reinforced concrete diaphragm resisting seismic forces is usually quite small relative to that of the vertical system it transfers force into. This is an example of a *rigid diaphragm*. Due to its in-plane stiffness it forces all vertical elements irrespective of their individual stiffness to deflect the same amount. The force each vertical element resists is therefore in proportion to its stiffness. A *flexible diaphragm* represents the other extreme. It is more flexible than the vertical structure beneath it. A common example of this is where relatively flexible timber diaphragms combine with stiff reinforced masonry or concrete walls. Since the diaphragm is too flexible to force all the walls to move together, each wall irrespective of its stiffness resists the inertia force only from the tributary area of floor connecting into it. So, depending upon the degree of diaphragm rigidity the forces resisted by individual vertical elements vary. This may require the structural engineer to fine-tune shear wall lengths or moment frame member dimensions, but the architectural implications of non-rigid diaphragms are usually minimal. Since small diaphragm deflections lead to less damage in a building, and it is always best to tie all building elements on any one level strongly to each other, diaphragm rigidity is definitely to be preferred.

A diaphragm can be considered analogous to that of a steel channel beam (Fig. 4.4). As its flanges provide bending strength, tension stress in one flange and compression in the other so *chords* – as they are called – do for diaphragms. Like the flanges of a beam, diaphragm chords should be continuous along the diaphragm length. Steel and wood diaphragms require the provision of specific chord members to carry the bending moment induced tensions and compressions. In concrete diaphragms these actions may be provided for by the simple addition of horizontal reinforcing steel along the diaphragm edges. However, a heavily laden concrete diaphragm might require specific chord members in the form of two longitudinal beams. Not only do they provide a location for accommodating the extra reinforcing steel but they prevent the compression edge of the diaphragm buckling.

The provision of diaphragm shear strength is also usually easy to provide. As the web of a steel beam withstands shear force, so does the horizontal plane of a diaphragm. Where a diaphragm is thin (constructed from plasterboard or plywood), the joists or rafters to which it is fixed provide out-of-plane stability to prevent it buckling under shear stress. Maximum shear force occurs at the diaphragm supports, namely adjacent to vertical elements such as structural walls into which the diaphragm transfers its shear. Strong connectors are required at these junctions. In reinforced concrete construction force

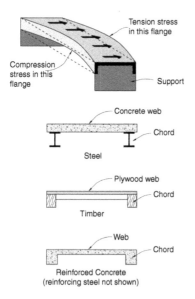

▲ **4.4**   A steel channel beam analogous to a diaphragm and cross-sections through diaphragms of different materials.

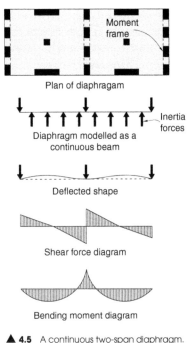

Plan of diaphragam

Diaphragm modelled as a
continuous beam

Inertia
forces

Deflected shape

Shear force diagram

Bending moment diagram

▲ **4.5**   A continuous two-span diaphragm.

transfer occurs through the concrete and reinforcing bars that tie horizontal and vertical elements together.

Just as buildings include both simply supported and continuous gravity beams, continuous diaphragms are commonly encountered. Fig. 4.5 shows a continuous diaphragm that spans between three supports (moment frames) in one direction. In the other direction the diaphragm is simply supported between two lines of shear walls. As discussed previously, the diaphragm can be modelled as a continuous beam with typical bending moment and shear force diagrams. A single chord of a continuous diaphragm therefore experiences tension and compression simultaneously in different sections along its length.

Before considering the construction materials used for diaphragms, mention must be made about the structural design of diaphragms given the philosophy of Capacity Design (Chapter 3). If designers have chosen to absorb seismic energy in structural fuses within primary vertical structural elements like structural walls, then all other structural elements including diaphragms must be designed strong enough to avoid damage. Damage should occur only within the specially detailed fuse regions.

Diaphragm materiality

The choice of diaphragm materiality depends upon the spans of the diaphragm and the intensity of inertia force to be resisted. While a house ceiling or roof diaphragm might span as little as 3 m and support the inertia force of light wood framing, a roof diaphragm over a sports stadium could span over 100 m and support heavy and high concrete walls. In the first example the diaphragm web might consist of single sheets of plasterboard or plywood nailed to wood framing and wooden chords. In the second, a braced steel diaphragm would be expected given that a wood solution would be too weak and concrete too heavy.

Where the roof and floors of a building are of reinforced concrete they, together with any perimeter beams, function as diaphragms. A cast-in-place floor slab is usually suitable as a diaphragm provided the structural engineer checks its strength for shear forces and bending moments. Where precast flooring is used, the precast units must be strongly connected so that they can work together to form an effective monolithic diaphragm. Usually a reinforced concrete topping slab provides a convenient connection method that also achieves a smooth and level

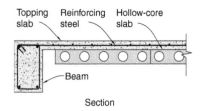

Topping slab   Reinforcing steel   Hollow-core slab

Beam

Section

▲ **4.6** Part section through a precast concrete and topping slab diaphragm.

floor surface (Fig. 4.6). Typically, the topping will be between 65 mm and 100 mm in thickness, contain reinforcing bars running in both directions and be cast over intentionally roughened precast concrete unit surfaces to achieve strong bonding between the fresh and hardened concrete.

Light-weight diaphragms consist of timber sheet products like plywood, particle board or wood cross-bracing (Fig. 4.7). In light steel construction, like that of the single-storey industrial building in Fig. 4.8, diaphragms usually take the form of horizontal cross-braced frames or trusses. The diaphragm consists of two horizontal trusses. They resist horizontal y direction forces and transfer them to the vertical cross-braced frames. Note that this diaphragm cannot resist nor transfer x direction forces. The diaphragm layout in Fig. 4.9 is suitable for that direction. Instead of a truss at each end of the building we could use a total of three or four trusses. This would allow truss member sizes to be reduced. If only one truss forms the x direction diaphragm its member sizes will be large since fewer members resist the same force. If this is a problem it can be partially alleviated by increasing the truss depth from one to two bays. Structural members like purlins running in the x direction need to be continuous and strongly connected to the horizontal trusses in order to transfer inertia forces into them. Roof purlins are sometimes doubled-up to fulfil this function.

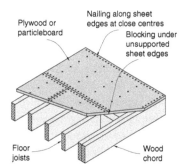

Plywood or particleboard

Nailing along sheet edges at close centres

Blocking under unsupported sheet edges

Floor joists   Wood chord

▲ **4.7** Section of a wooden floor diaphragm.

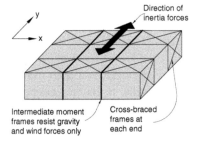

Direction of inertia forces

Intermediate moment frames resist gravity and wind forces only

Cross-braced frames at each end

▲ **4.8** Structural layout of a single-storey light-industrial steel building.

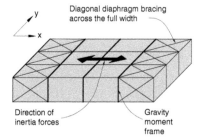

Diagonal diaphragm bracing across the full width

Direction of inertia forces

Gravity moment frame

▲ **4.9** Braced diaphragm and cross-bracing for x direction forces.

Member continuity and strength is also required in the y direction to form the truss chords. As well as functioning as the beams of gravity moment frames these horizontal members experience additional tension and compression stress during an earthquake when they take on a second structural role as diaphragm truss chords.

The final step of diaphragm design involves combining the requirements of the previous two diagrams (Fig. 4.10). Two of many possible

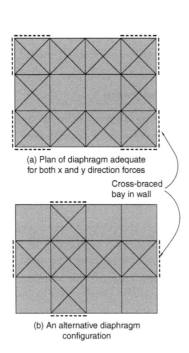

(a) Plan of diaphragm adequate
for both x and y direction forces

Cross-braced
bay in wall

(b) An alternative diaphragm
configuration

▲ **4.10** Diaphragm and wall bracing
configurations suitable for both x and y
direction forces.

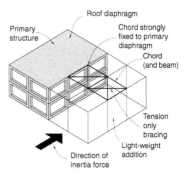

▲ **4.11** A secondary diaphragm
cantilevers horizontally from the primary
structure to avoid vertical end-wall cross
bracing.

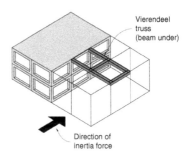

▲ **4.12** A vierendeel truss as a
cantilevered diaphragm.

diaphragm configurations are illustrated. Now inertia forces in *any* direction have an identified force path. Depending on the levels of force these cross-braced diaphragms can consist of tension-only bracing or tension and compression bracing. Although tension-only bracing requires twice as many diagonal members they are far smaller in cross-section than any designed to resist compression.

Tension-only diaphragm bracing is also useful where connecting a light-weight glazed addition to a heavier and stronger primary structure (Fig. 4.11). Here the roof bracing, designed for wind as well as seismic force spans the width of the lean-to. As there is no vertical bracing within the end-wall of the addition the roof diaphragm needs to cantilever horizontally from the primary roof diaphragm.

Regarding the choice of diaphragm configuration the two types already illustrated are either opaque, as in the case of sheet-based wood or concrete diaphragms, or essentially transparent. Daylight penetrates trussed diaphragms with transparent roof cladding. If the exposed diagonals of normal truss diaphragms are unacceptable architecturally, vierendeel trusses are an option, albeit expensive (Fig. 4.12).

Architects enjoy considerable freedom when configuring roof diaphragms. Ideally, a diaphragm should provide the most direct force path for inertia forces into vertical bracing elements yet it can sometimes provide opportunities to strengthen the expression of the architectural design concept or idea (Fig. 4.13).[1]

Diaphragm penetrations

Architects penetrate diaphragms for a variety of purposes. One of the most common reasons is to provide vertical circulation for stairs and elevators. Interior design aspirations might also culminate in larger openings, perhaps creating localized double-height volumes. Services penetrations are also common but their lesser plan dimensions do not usually adversely affect diaphragm performance.

Large penetrations should be located where they do not jeopardize the ability of a diaphragm to transfer its forces to the vertical structure that stabilizes the building. The seismic force path must not be interrupted. Penetrations are, therefore, best located in areas of either low bending moment or shear stress as indicated by the shapes of the bending moment and shear force diagrams (Fig. 4.14). The penetration in Fig. 4.14(a) cuts through a chord. This is like cutting a notch in the flange of a steel I-beam – a recipe for disaster. However, the choice of penetration location is structurally acceptable if the edge beam continues through the penetration to restore continuity of the diaphragm chord (Fig. 4.14(b)). Another poorly conceived penetration is shown in Fig. 4.14(c). Its placement coincides with the location of maximum shear force (Fig. 4.3). Since only narrow areas of diaphragm connect into the structural wall shear failure is probable in these weakened regions. The length of penetration needs to be reduced or the diaphragm thickened and far more heavily reinforced in those areas to avoid that failure mode. In Fig. 4.14(d) the penetration in the middle of the web is safely away from the chords that act in tension and compression and in an area where the shear force is least. For suggestions on how to design other potentially serious diaphragm penetrations such as long light-slots and other interruptions refer to Chapter 9.

TRANSFER DIAPHRAGMS

The diaphragms discussed previously are sometimes termed *simple* diaphragms. They resist the inertia forces from their own mass and those of elements like beams and walls attached to them. *Transfer* diaphragms resist the same forces but in addition they transfer horizontal forces

▲ **4.13** Horizontal forces from light-weight construction to the left are transferred through roof and first floor braced diaphragms to concrete masonry shear walls. Educational building, London.

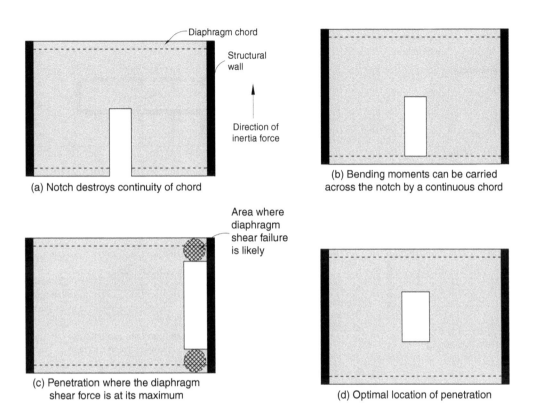

(a) Notch destroys continuity of chord

(b) Bending moments can be carried across the notch by a continuous chord

(c) Penetration where the diaphragm shear force is at its maximum

(d) Optimal location of penetration

▲ **4.14** Diaphragm penetrations in various locations.

from one vertical bracing system above to another that is offset horizontally below. Transfer diaphragms are usually far more heavily stressed than simple diaphragms and consequently need to be considerably stronger and may not be able to accommodate large penetrations.

In the building shown in Fig. 4.15(a), y direction forces are resisted by structural walls at each end. One wall at first floor level is discontinuous. It does not continue directly from the roof to the foundations. Therefore, the horizontal forces at the bottom of that wall must be transferred through the first floor slab functioning as a transfer diaphragm into the offset wall at ground floor. The forces acting on the transfer diaphragm are shown in Fig. 4.15(c). If the floor is reinforced concrete, at the very least it may require extra reinforcement but additional thickening may also be needed. In this building the transfer diaphragm transmits all the horizontal shear forces from the wall above to the wall beneath. Since the diaphragm is incapable of resisting the overturning moment from the wall above, two columns, one

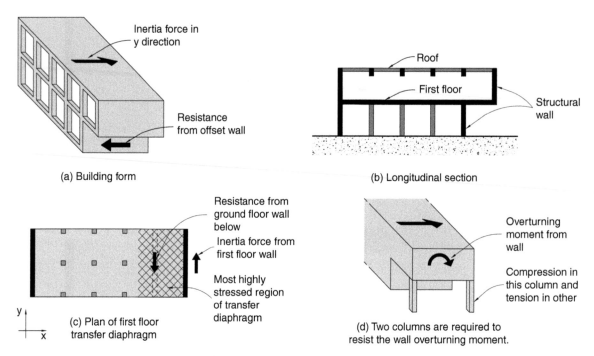

Inertia force in y direction

Resistance from offset wall

(a) Building form

Roof

First floor

Structural wall

(b) Longitudinal section

Resistance from ground floor wall below

Inertia force from first floor wall

Most highly stressed region of transfer diaphragm

(c) Plan of first floor transfer diaphragm

Overturning moment from wall

Compression in this column and tension in other

(d) Two columns are required to resist the wall overturning moment.

▲ **4.15** A transfer diaphragm provides a force path for horizontal shear forces from the upper wall to the offset ground floor wall.

under each end of the wall, are required to resist the tension and compression forces from that moment and thereby stabilize the wall (Fig. 4.15(d)). If these columns are architecturally unacceptable the only other option is to provide deep transfer beams (Chapter 9).

Due to their less direct force paths that potentially lead to increased seismic damage, transfer diaphragms and any associated vertical structure need to be designed very carefully. Capacity Design procedures must be followed to prevent undesirable failure mechanisms. Although designers prefer to avoid transfer diaphragms, sometimes they are unavoidable and cause significant design complications (Fig. 4.16).

Bond beams

Bond beams, introduced in Chapter 2, offer another approach to resisting horizontal inertia forces and transferring them sideways to bracing elements (see Fig. 2.19). In the absence of a floor or roof diaphragm a bond beam can span horizontally between lines of vertical bracing elements like shear walls. Although designers use bond beams frequently in masonry construction the same principle can be applied

▲ **4.16** A hidden eccentrically braced frame immediately behind precast concrete panels is offset from the ground floor braced frame. A short transfer diaphragm moves horizontal shear from the outer frame to the inner ground floor frame (Left). To resist the overturning-induced compressions and tensions from columns above, the first floor cantilever beams are very deep (Right). Apartment building, Wellington.

to all construction types and materials. Where the distance between bracing walls requires a relatively deep or wide horizontal beam, a truss might be more suitable.

Imagine designing what is essentially a box-shaped form shown in Fig. 4.17(a). If a desire for a very transparent roof precludes a normal braced diaphragm, bond beams may be a solution for effectively creating a flexible roof diaphragm. Another solution is to cantilever the walls vertically from their foundations but very thick walls or walls with deep ribs at, say, 4 to 5 m centres, may not be acceptable architecturally (Fig. 4.17(b)). If bond beams are chosen, the wall thickness must be sufficient to span vertically between the roof level bond beam and the foundations. The bond beams are then designed to resist half of the out-of-plane forces on a wall and to transfer them to the two walls parallel to the direction of seismic force. By virtue of spanning horizontally rather than vertically, the orientation of bond beams is unusual. It is like taking a normally proportioned beam and rotating it 90 degrees to act horizontally. The relationship of the beam to the wall can be varied as shown in Fig. 4.17(c).

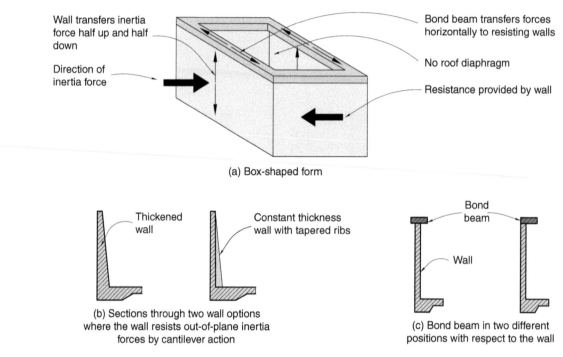

Wall transfers inertia force half up and half down

Direction of inertia force

Bond beam transfers forces horizontally to resisting walls

No roof diaphragm

Resistance provided by wall

(a) Box-shaped form

Thickened wall

Constant thickness wall with tapered ribs

Bond beam

Wall

(b) Sections through two wall options where the wall resists out-of-plane inertia forces by cantilever action

(c) Bond beam in two different positions with respect to the wall

▲ **4.17** Vertical cantilever and bond beam options for resisting out-of-plane inertia forces on the walls of a box-shaped building.

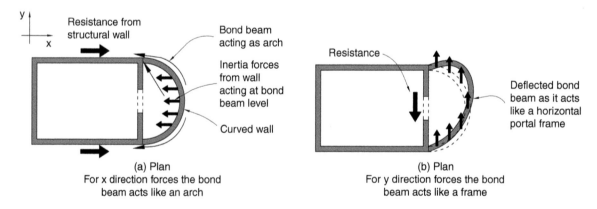

y

x

Resistance from structural wall

Bond beam acting as arch

Inertia forces from wall acting at bond beam level

Curved wall

Resistance

Deflected bond beam as it acts like a horizontal portal frame

(a) Plan
For x direction forces the bond beam acts like an arch

(b) Plan
For y direction forces the bond beam acts like a frame

▲ **4.18** A curved bond beam behaves differently depending on the direction of horizontal force.

If an exterior wall is curved, a bond beam can use the rounded geometry to transfer forces primarily by arch-action (Fig. 4.18(a)). In the orthogonal direction the bond beam-cum-arch would be designed as a rounded portal frame (Fig. 4.18(b)).

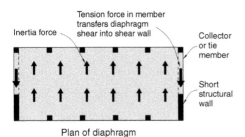

Plan of diaphragm

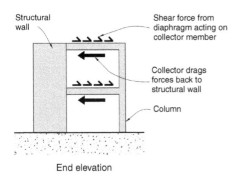

End elevation

▲ **4.19** Collector or tie members transfer diaphragm shear forces into vertical structure.

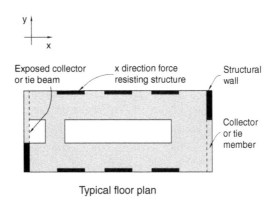

Typical floor plan

▲ **4.20** A car parking building where collector members are needed to transfer y direction shear forces into the two shear walls.

▲ **4.21** Collector or tie beams connect into the end of a shear wall. They transfer inertia forces from a diaphragm on the right into the shear wall. Car parking building, Wellington.

COLLECTORS AND TIES

Where the length of a bracing element – such as a shear wall – is short in plan with respect to the width of the diaphragm transferring forces into it, the interface between the horizontal and vertical element may be too weak to transfer the forces between them. In this case, a 'collector' or tie member is required. It collects forces from the diaphragm acting in either tension or compression, depending upon the direction of force at that instant of time and transfers them into the wall (Fig. 4.19). Collector members need to be strongly joined to bracing elements, in this case shear walls. In reinforced concrete construction a collector may consist of merely a few additional reinforcing bars embedded in the floor slab and well anchored into the vertical structure. Where large forces are to be transferred, or where a gap must be crossed, special collectors or tie beams are required. Strong in tension and compression they contain enough longitudinal steel to physically tie one part of a structure to another, as in the car parking building shown in Figs 4.20 and 4.21.

NOTE

1 For an example of an unusually configured diaphragm contributing architecturally see Balmond, C. (2002). *Informal*. Prestel, p. 64.

5 VERTICAL STRUCTURE

INTRODUCTION

Chapter 4 explained the importance of horizontal structure in the form of floor and roof diaphragms, bond beams and collector or tie members. The role of all these members is to resist inertia forces and then transfer them horizontally into vertical structure. Vertical structure resists those forces and, in next stage of the force path, transfers them downwards through the strength of its members be they walls, columns or diagonal braces and into the foundations.

The vertical structure required for seismic resistance is often very different from that resisting gravity forces. In its two most simple forms, a gravity resisting structure consists of post-and-beam or load-bearing wall construction (Fig. 5.1). The vertical elements of both systems support their load by compression alone. They require sufficient cross-sectional dimensions only to prevent buckling. Consequently, gravity frameworks usually offer little or no resistance to horizontal forces. Post-and-beam structures with pins top and bottom of the columns are completely unstable. Frames designed to resist gravity forces only, where joints between columns and beams are rigid enough to form moment frames, may be more stable against horizontal forces depending on the slenderness of their columns. As for load-bearing walls, they are usually weak at their base with respect to out-of-plane forces, overturning easily when loaded in that direction. Their length does offer bracing potential when they function as shear walls to resist in-plane forces, but this inherent capacity is realized only by intentional structural design and detailing.

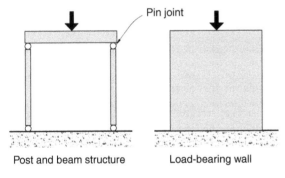

Pin joint

Post and beam structure Load-bearing wall

Stable under vertical load

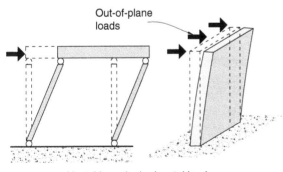

Out-of-plane loads

Unstable under horizontal load

▲ **5.1** Instability of two gravity force resisting structures against horizontal forces.

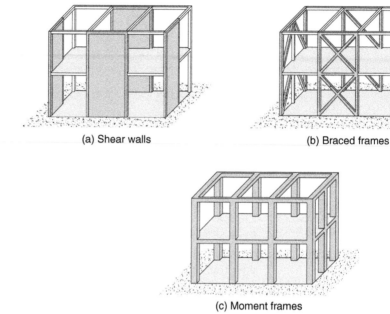

(a) Shear walls

(b) Braced frames

(c) Moment frames

▲ **5.2** The three basic seismic force resisting systems.

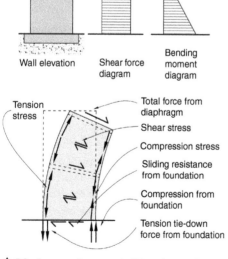

Forces acting on wall from diaphragm

Wall elevation

Shear force diagram

Bending moment diagram

Tension stress

Total force from diaphragm

Shear stress

Compression stress

Sliding resistance from foundation

Compression from foundation

Tension tie-down force from foundation

▲ **5.3** Forces acting on and within a shear wall.

Overview of seismic force resisting systems

The choice of vertical structural systems to resist horizontal seismic forces is quite limited. The three most common systems illustrated in Fig. 5.2 comprise:

- Shear walls
- Braced frames, and
- Moment frames.

Although each system is also capable of resisting gravity forces, its primary function, in the context of this book, is to provide horizontal resistance. Acting as a vertical cantilever each system is grounded by its foundations and rises up a building to receive horizontal forces from each floor and roof diaphragm (Fig. 5.3). It experiences horizontal shear forces and bending moments which give rise to vertical compression and tension forces concentrated at each end of the wall or frame. Listed in order of decreasing stiffness, shear walls, braced frames and moment frames can be thought of as vertical cantilever structural walls with different degrees and geometries of penetrations to realize cross-braced and

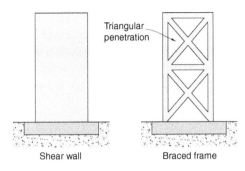

Shear wall   Braced frame

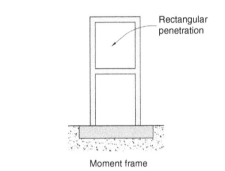

Moment frame

▲ **5.4** Penetrations of a shear wall lead to braced or moment frame systems.

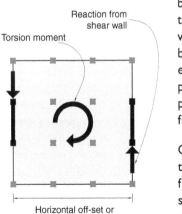

▲ **5.5** Seismic force in the y direction is resisted by two shear walls. Because of the off-set or lever-arm between them they also resist torsion. Structure acting in the x direction is not shown.

moment frame forms (Fig. 5.4). A brief overview of each of these three structural systems precedes more in-depth discussions that follow.

A shear wall resists horizontal forces acting in its plane by virtue of its length and to a lesser extent its thickness. Its strength and stiffness against horizontal forces usually requires a rigid connection to its foundations. Like any of the three systems providing seismic resistance it should be continuous from foundation to roof. The material of a wall can consist of any recognized structural material provided it has sufficient strength to resist the forces that the diaphragms transfer into it. Usually the wall material will be as strong if not stronger than that of the diaphragms.

Braced frames come in a variety of configurations. In their most basic form they consist of posts, beams and one or two diagonal bracing members per storey that generally fully triangulate the structure. Essentially, braced frames are vertical trusses. All their joints can be pinned. Depending on the type of frame and the cross-sectional dimensions of its members, diagonal members resist tension forces only or both tension and compression.

By contrast, a moment frame requires rigid connectivity between beams and columns. Rigid and strong joints enable bending moments to be transferred between adjacent column and beam members without any relative rotation. Horizontal forces are resisted mainly by bending and shear forces in the beams and columns. Due to the ever-present overturning moment causing toppling that acts upon any primary vertical structure, end columns of a frame experience compression and tension forces in addition to the gravity forces present. A frame can consist of any number of bays and storeys.

One of these three systems must be present in *each* orthogonal direction when considering the plan of a building. Then earthquake attack from *any* direction is resisted (see Fig. 2.13). Strive for symmetry of structural layout to reduce torsion. Ideally the lengths and thicknesses of shear walls acting in one direction should be similar. Where two or more elements of any system, say shear walls, are off-set in plan creating a lever-arm between them they can resist both horizontal forces and any in-plan torsion (Fig. 5.5) as discussed further in Chapter 9. The numbers of structural elements necessary in each direction depends upon a number of factors including site seismicity, the weight of the building, and its height. Architects also decide upon the numbers of elements to meet their design criteria bearing in mind the fewer the elements the larger the force each one is required to carry. It is also necessary to integrate the seismic force resisting systems with the gravity resisting structure and with the architectural concept and program or design brief.

It is recommended that just one structural system only is used in each orthogonal direction. It is permissible, and in fact sometimes desirable, to place different structural systems in parallel but because mixed-systems — as they are known — lead to more complex force paths, they are best avoided until those implications are studied later in this chapter.

SHEAR WALLS

Shear walls are structural walls designed to resist horizontal force. The term 'shear wall' originally referred to a wall that had either failed or was expected to fail in shear during a damaging quake. Now that a primary objective for contemporary structure is to avoid shear failure, 'shear wall' is a somewhat inappropriate description for a modern well-designed wall. However, given the term's international popularity, its on-going usage is justified by appreciating that a shear wall is designed to primarily withstand horizontal shear forces. In the process, of course, it experiences bending moments and a tendency to overturn (see Fig. 5.3).

Of all seismic resistant structural systems, reinforced concrete shear walls have the best track record. During past earthquakes even buildings with walls not specially detailed for seismic performance, but with sufficient well-distributed reinforcement, have saved buildings from collapse. The success of shear walls in resisting strong earthquake shaking has led some leading structural engineers to recommend them, at least for reinforced concrete construction. For example, Mark Fintel a noted US structural engineer who studied the seismic performance of shear wall buildings over a thirty-year period, concludes: 'We cannot afford to build concrete buildings meant to resist severe earthquakes without shear walls.'[1]

Shear wall buildings are the best choice in earthquake-prone countries. Reinforced concrete shear walls are relatively easy to construct because their reinforcing details are straight-forward, at least when compared to those of moment frames. Their inherent stiffness minimizes horizontal interstorey deflections and consequently earthquake

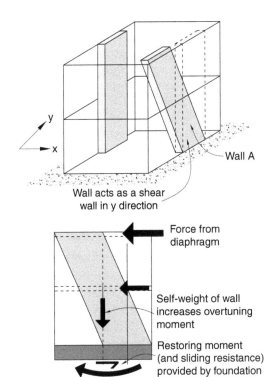

Wall acts as a shear
wall in y direction

Force from
diaphragm

Self-weight of wall
increases overtuning
moment

Restoring moment
(and sliding resistance)
provided by foundation

Elevation Wall A

▲ **5.6** The shear wall sloping in both directions is effective
in the y direction but needs support from x direction
structure (not shown).

▲ **5.7** Free-standing shear walls prior to incorporation into
surrounding structure highlight their vulnerability to out-of-plane
forces. University building, Vancouver.

damage to structural and non-structural elements. The challenge of integrating shear walls with architectural form and planning and their increased cost with respect to inferior masonry walls prevent them being more widely adopted internationally.

Given the excellent seismic credentials of shear walls as a source of seismic resistance, architects should always try to use them in the first instance. Approach each design project with a view to walls resisting seismic forces in both directions. Often walls required primarily for architectural reasons, such as exterior cladding or enclosing stairwells and so on can function as shear walls. However, shear walls need to be long enough in plan so as to attract and resist seismic forces before other structural elements in a building like columns and beams suffer damage.

Where intended to function as a seismic resisting system, a shear wall is usually continuous up the height of a building from its foundations to the uppermost diaphragm at roof level. If a shear wall is discontinuous through a storey or has one or more off-sets in plan up its height it is possibly fatally flawed, as discussed in Chapter 8. But it is permissible to slope a shear wall out-of-plane and even moderately in-plane without significantly affecting its structural performance. The additional gravity bending moments that arise from an in-plane lean require add-itional wall and foundation strength (Fig. 5.6). Although most shear walls are rectangular in plan many other plan shapes such as a gentle curve or C, L and I-shapes are usually structurally feasible.

Finally, it must be reiterated that a shear wall is *effective in the direction of its length only*. When laden at right angles to its length – that is, when subject to out-of-plane forces – a typical shear wall is very weak and flexible and likely to collapse unless restrained at every one or two storeys up its height by diaphragms. In summary, when acting in-plane shear walls support but where loaded out-of-plane they require support (Fig. 5.7).

▲ **5.8** A reinforced concrete shear wall strengthens an existing building. Note how the number of penetrations per storey increases with height. Telephone exchange, Wellington.

Structural requirements

Figure 5.2(a) illustrates a simple building where seismic forces are resisted by four shear walls, two acting in each orthogonal direction. Inertia forces are transferred into the walls by diaphragms at each level. Shear forces and bending moments in a wall increase with the distance from the roof and reach their maximum values at the foundations (see Fig. 5.3). A structurally adequate wall possesses sufficient strength to resist both shear forces and bending moments. Shear force is resisted by the web area of a wall; that is, its length times its thickness. The shear strength of reinforced concrete walls is provided by a combination of the concrete strength and horizontal reinforcing bars that act as shear reinforcement. Penetrations for windows and doors can reduce the shear strength of a wall significantly. For that reason, where a wall is highly stressed, particularly near its base, such large penetrations might not be possible (Fig. 5.8).

Bending moments cause tension and compression forces at each end of a wall. Some walls are provided with 'chords' or thickenings at their ends specifically to provide the cross-sectional area necessary to accommodate the vertical tension reinforcement and also to prevent the ends of walls buckling in compression. For the lowest storey of ductile shear walls where plastic hinges form, the ratio of the clear interstorey height to chord or wall thickness, whichever is greater, should be less than or equal to 16.[2] Where the maximum bending moment is low with respect to wall length, chords may not be necessary for concrete or masonry construction. Often sufficient bending strength can be achieved with adequate vertical reinforcing placed within the wall and provided that buckling is suppressed by provision of sufficient wall thickness.

Shear forces and bending moments, as well as gravity compression forces acting on a wall, are ultimately resisted by the foundations. Potential wall sliding due to shear force is avoided by friction between the base of the foundations and the ground, and horizontally induced soil pressure. Wall over-turning or toppling is counteracted by one or more of the following mechanisms: the inherent stability of a wall due to its length and the gravity force acting upon it including its own weight, a foundation beam with a footing to increase the lever-arm between the line of gravity force through the wall and the centre of soil-pressure, and tension piles (Fig. 5.9). On some sites rock or ground anchors, well protected from corrosion, provide the necessary tension forces to resist overturning.

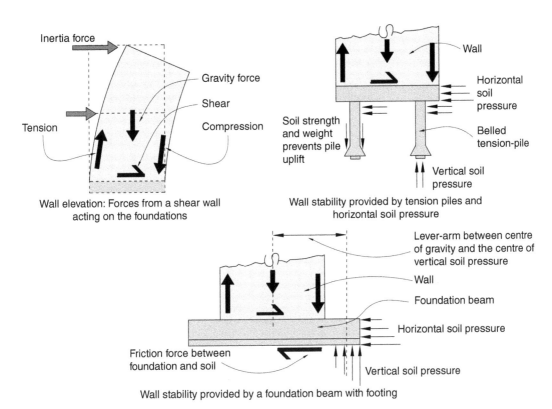

Wall elevation: Forces from a shear wall acting on the foundations

Wall stability provided by tension piles and horizontal soil pressure

Wall stability provided by a foundation beam with footing

▲ **5.9** How foundations resist overturning moments and shear forces from a vertical structure, in this case a shear wall.

Shear wall material and height

As noted previously, a shear wall can be constructed from any structural material. Provided its material strength is not significantly weaker than that of the surrounding structure its material of construction is likely to be structurally viable. Choice of shear wall materiality is also influenced by building height and other constraints as summarized in Table 5.1. Readers with the frequently asked question: 'How many shear walls of what thickness and length are necessary for my project?' must wait until the following chapter for an answer.

While still on the subject of shear wall materiality, unreinforced masonry warrants further discussion due to its extensive usage internationally. Its different manifestations in buildings include completely unreinforced masonry construction as in Fig. 5.14, as infill panels between reinforced concrete or steel moment frames, as confined masonry walls, or as non-structural partition walls. As discussed in Chapter 10 and illustrated in Fig. 10.4(a), a masonry infill panel within a moment frame functions as a diagonal bracing strut when a combined frame and infill system is loaded horizontally. Provided the frames and infills satisfy the requirements in Table 5.2 they *may* contribute positively to the seismic resistance of a building. In many situations, however, some of the listed requirements are architecturally unacceptable. In such cases, infills should be physically separated

▼ **5.1**   Common shear wall materials, their typical ranges of height and general comments

Material	Typical height range (storeys)	Comments
Steel	4–8	Steel plate walls have occasionally strengthened existing buildings but otherwise are rarely used (Fig. 5.10).
Reinforced concrete	1–20	The most reliable material for medium to heavy construction. Its high level of ductility enables designers to achieve the shortest wall lengths. Reinforced concrete walls are usually cast in-place (Fig. 5.11). Precast concrete panels including tilt-up panels typically one- to three-storeys high, and other panels strongly connected together can also form monolithic reinforced concrete walls (Fig. 5.12).
Reinforced masonry	1–6	Popular where long wall lengths are achievable such as adjacent to boundaries. As the strength of masonry units and grout infill reduces, required wall lengths increase (Fig. 5.13). Construction quality control requires special attention.
Unreinforced masonry	1–4	Due to its inherent lack of ductility this material is prohibited in some countries and in others permitted only in less seismically active areas (Fig. 5.14).
Confined masonry	1–6	'Confined' masonry walls incorporate panels of unreinforced masonry within a reinforced concrete beam and column frame. They are discussed below.
Wood	1–5	Suitable for light-weight construction. Because of its relatively low strength and stiffness as compared to reinforced concrete or steel construction several long or many shorter walls may be required to resist seismic forces. Particularly appropriate for apartments and similar buildings where function dictates a cellular layout. Plywood fixed to wood framing often forms the web. Chords can be from sawn or glue-laminated wood or even a more highly engineered wood product like laminated veneer lumber (LVL). Where horizontal deflections are critical steel chord members may be specified (Fig. 5.15). Gypsum plasterboard or equivalent sheets are nailed or screwed to wood or steel framing to form shear walls in light-weight framed construction. A popular and cost-effective material for domestic construction in New Zealand, but the wall lengths required are longer than those using a stronger wood-based product such as plywood.

▲ **5.10** Steel shear walls during construction. Hospital, Portland, Oregon.
(Reproduced with permission from KPFF).

▲ **5.11** Reinforcing steel for cast-in-place concrete shear walls. High-rise office building, San Francisco.

▲ **5.12** A shear wall formed from numerous precast panels strongly connected vertically and horizontally. Apartment building, Wellington.

▲ **5.13** A reinforced concrete masonry wall with attached columns under construction. Horizontal reinforcing, placed every third masonry unit is not visible. Office building, Wellington.

▲ **5.14** An unreinforced masonry wall under construction. Kanpur, India.

▲ **5.15** A two-storey wood shear wall with steel chords. Wanganui, New Zealand.

▼ **5.2** Recommendations for the safe use of masonry infill walls within reinforced concrete or steel moment frames

Discipline impacted by the requirement	Requirements of masonry infills and moment frames
Architectural	Infills reasonably symmetrically located in plan to avoid excessive torsion.
	Infills continuous up the height of the building beginning from the foundations.
	If infills are penetrated openings should be the same size and in the same position.
Structural	Infills should be physically connected to columns or beams with reinforcement to prevent the masonry collapsing under out-of-plane forces.
	Moment frames are designed to resist horizontal forces without the assistance of infills.
	Special ductile detailing at beam-column junctions and at the tops of each column adjacent to an infill to prevent the infill diagonal compression strut causing premature column damage.

from the structure or else the structure made so stiff and strong as not to be adversely affected by them. The positive and negative contributions of masonry infills to the sound seismic performance of buildings is a topic of on-going research and is discussed in more detail in Chapter 10.

The difference between an 'infill' and 'confined' masonry wall panel is that confined masonry *is* intended to play a structural role.[3] Masonry is confined between columns and beams that are cast *after* the masonry is laid thereby thoroughly integrating it structurally and allowing diagonal struts to form within the masonry under horizontal forces (Fig. 5.16). This means large penetrations like doors that prevent the formation of diagonal struts are not permitted in confined masonry panels that are expected to provide seismic resistance. Confined masonry panels are both shear walls and gravity load-bearing walls. Beam and column confining members are not designed as elements of moment frames but rather as chords and ties of braced frames. The masonry acts as diagonal compression members (Fig. 5.17).

While confined masonry construction enables column and beam dimensions to be minimized due to an absence of seismically-induced bending moments significant architectural limitations beyond those of Table 5.2 are necessary. Because the seismic strength of a confined masonry building is reliant upon the masonry panels enough panels of sufficiently thick masonry need to be provided in each orthogonal direction to provide the necessary horizontal strength. In the example in Fig. 5.18 the cross-sectional area of structural walls along each axis is 8 per cent of the gross floor area. This compares to a value of

▲ **5.16** Confined masonry construction. Columns and beams are cast after laying the walls. Only walls without large penetrations and are continuous up the height of the building act as shear walls. Lima.
(Reproduced with permission from Ángel San Bartolomé)

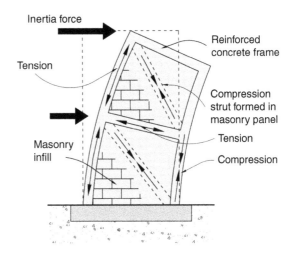

Elevation of a confined masonry wall showing internal 'truss action' forces

▲ **5.17** The forces within a confined masonry wall.

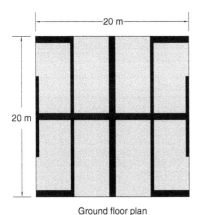

Ground floor plan

▲ **5.18** Plan of a four-storey confined masonry wall and concrete floor building in the most seismically active zone of Indonesia. The significant lengths and thicknesses of walls drawn to scale are required to satisfy structural requirements. All walls are constructed from brick masonry and are not penetrated.

3 per cent suggested for Iranian school construction up to three storeys high.[4] Since the masonry of confined masonry construction can withstand only modest diagonal compression stress depending upon the quality of masonry units and the mortar, the height of this type of construction should be limited. One report, based upon Indonesian conditions, recommends these buildings not exceed four storeys in height.[5]

As for infill walls, researchers continue to develop seismic resistant guidelines for this common construction system. Infill walls are discussed further and in detail in Chapter 10.

Shear wall ductility

Assume an architect has chosen shear walls as the seismic resisting system for at least one orthogonal direction of a building. If short rather than long shear walls are preferred – perhaps to ease architectural planning constraints – ductile walls with their inherently lower design forces and shorter lengths are recommended.

The question then arises: How to design a *ductile* shear wall? For an answer we return to Chapter 3. The Capacity Design approach must be applied. In summary, a ductile overload mechanism is identified and all possible brittle failure modes are suppressed. Table 5.3 compares

▼ **5.3** Ductility of shear walls constructed of various materials

Shear wall material	Typical degree of ductility	How ductility is achieved
Steel	Medium to high	Yielding diagonal tension zones between opposite corners of the steel panels absorb seismic energy. This ductile elongation occurs before damage to the many bolts connecting the steel panels to surrounding frame members.
Reinforced concrete	High	Once bending moments exceed the bending strength of a wall a structural fuse or plastic hinge forms at its base. Vertical reinforcement in the chords at each end of the wall yields in tension and compression. The wall shear strength and the strength of its foundations are designed stronger to preclude premature failure. In the structural fuse region, typically the lesser of a storey-height or the wall length above foundation level door or window penetrations are inadvisable due to high stresses and the need to confine the concrete with horizontal ties (Fig. 5.19).
Reinforced masonry	Medium to low	A similar but less ductile performance as compared to a reinforced concrete wall can be achieved. It is impractical to place confining steel into narrow masonry units but short and thin steel plates inserted into the mortar joints at the ends of a wall prevent premature crushing of masonry units.
Confined masonry	Low	Diagonal cracking within the masonry panels may be followed by damage to the tops of columns from the diagonal compression struts. The presence of reinforcing steel ensures some ductility.
Unreinforced masonry	None	No ductility is expected.
Wood	Medium to high	The chosen ductile mechanism is usually the bending deformations in the hundreds of nails between a plywood web and the supporting framing. Wall chords, their connections to the foundations, the foundations and the plywood itself are designed not to fail before the nails absorb earthquake energy by yielding as they bend to-and-fro (Fig. 5.20). Gypsum plasterboard on wood framing is not usually rigorously designed for ductility. During a design-level earthquake, nails distort and damage the plaster in their vicinity and the holding-down fixings between the wall chords and foundations might be damaged. A low to medium level of ductility is achievable.

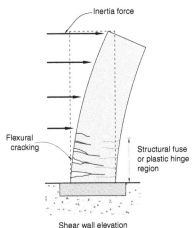

Shear wall elevation

▲ **5.19** Structural fuse region at the base of a ductile shear wall. The vertical reinforcing yields and absorbs earthquake energy.

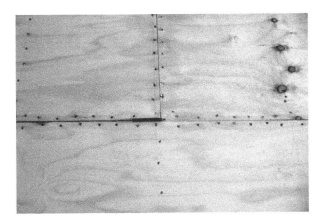

▲ **5.20** The plywood web of a wood shear wall is fixed to framing with closely-spaced nails that transfer forces between sheets and yield in bending to act as structural fuses. School building, Wellington.

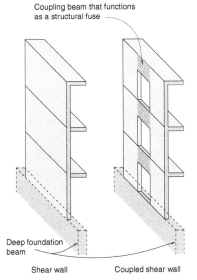

Shear wall   Coupled shear wall

▲ **5.21** Comparison between a shear wall and a coupled shear wall.

the ductility of shear walls with different materiality and notes how ductility is achieved. Unlike moment frames discussed in a following section the ductile design requirements of shear walls, apart from restrictions on wall penetrations in the structural fuse region, have few architectural implications.

Coupled reinforced concrete walls

A coupled shear wall in terms of its strength and stiffness can be imagined as lying halfway between a large unpenetrated wall and two individual walls whose combined lengths equal that of the single wall (Fig. 5.21). Envisage a coupled shear wall as two or more walls coupled or connected by deep coupling beams, or alternatively as a highly-penetrated single wall. For maximum structural effectiveness the depth of the coupling beams is normally greater than half their clear span, much deeper than conventional beams. Coupling beams force the walls to work together to resist seismic forces. Where walls are coupled by shallow beams or merely tied together by floor slabs they can be considered as independent walls that are considerably more flexible and weaker than if joined by coupling beams.

The architecturally defining feature of coupled shear walls is their squat coupling beams usually at each storey. They are readily identified during construction by their characteristic diagonal reinforcing bars which provide the system with its high level of ductility (Fig. 5.22). Where coupled reinforced concrete walls are designed in accordance with the Capacity Design approach structural fuses form in the coupling beams

▲ **5.22** Diagonal reinforcement in a coupling beam. Office building, San Francisco.

and at the base of each wall before any shear or foundation damage occurs. Although fuses suffer structural damage when they absorb
earthquake energy, thanks to the concrete confinement provided by
closely-spaced ties they maintain their strength and are repairable after
a damaging quake.

BRACED FRAMES

Introduction

Although the braced frame is the least common vertical seismic resisting system from an international perspective it is used extensively
in some countries and offers architects an alternative to walls and
moment frames. As steel-framed buildings become increasingly popular in more industrialized countries braced frame usage will continue
to rise. However, in countries where structural steel is less affordable
and the quality of welded connections less dependable, braced frames
will remain relatively under-utilized.

Types of braced frames

Often concealed within the walls of building cores braced frames cantilever vertically from their foundations to resist the seismic forces
transferred from diaphragms. The basic types of braced frames are
illustrated in Fig. 5.23.

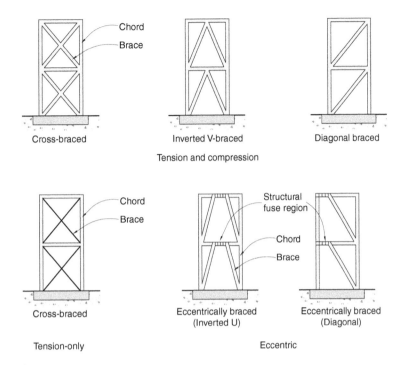

▲ **5.23** Common types of braced frames.

The diagonal members for all frame types except for tension-only braced frames are designed to resist tension and compression. Member cross-sections are therefore quite large to resist buckling. For multi-storey steel construction typically steel tubes or steel universal column sections are used for bracing members subject to compression.

Tension-only bracing is particularly common in low-rise and light-industrial buildings. It is usually very cost-effective since it utilizes steel in its most efficient structural mode – tension. The bracing members are usually very slender, such as steel rods or flats, so their compressive strengths are negligible. Depending on the building weight and the number of braced frames bracing member cross-section diameters can be as little as 20 to 30 mm. The advantage of this type of bracing from an architectural perspective is its economy and transparency. Vertical tension-only bracing is often used in conjunction with similar roof diaphragm bracing (Fig. 5.24).

▲ **5.24** Tension-only bracing. Light industrial building, Wellington.

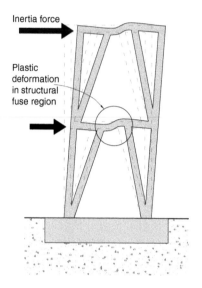

Inertia force

Plastic deformation in structural fuse region

▲ **5.25** The deformation of an eccentrically braced frame showing the distortion of the structural fuses.

▲ **5.26** An eccentrically braced frame with stiffener plates and fly-braces to stabilize the beam bottom flange in the structural fuse region. Apartment building, Wellington.

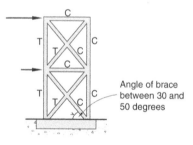

Angle of brace between 30 and 50 degrees

T = Tension
C = Compression

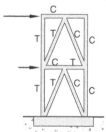

▲ **5.27** Two tension and compression fully triangulated frames. The types of axial force are indicated.

Eccentrically braced frames are the most ductile of all braced frames. The eccentricity between the inclined braces or between columns and braces ignores the centuries-old best-practice of concentric connections. Consequently, severe seismic bending moments and shear forces form in the beam fuse region located between braces. Plastic bending and shear deformation of the steel beam absorbs earthquake energy (Fig. 5.25). Special welded stiffener plates in the fuse region prevent the beam web from buckling (Fig. 5.26). The beams of eccentrically braced frames may be pin-joined to columns but they must be continuous over the braces. The floor slab supported by the beam will suffer damage unless it is separated from the fuse region and its potentially severe distortions.

Structural requirements

As mentioned in the introduction to this chapter, braced frames are essentially pin-jointed structures. Most of their horizontal load-carrying capability is achieved by their members working in either pure compression or tension. Apart from the beams of eccentrically braced frames individual members do not resist significant bending moments. The ability of braced frames to function as vertical trusses lies in their triangulation (Fig. 5.27). Clearly, if any members are required to resist

▲ **5.28**   Sub-floor wood bracing of a wood framed house. Wellington.

▲ **5.29**   Reinforced concrete tension and compression bracing. A high level of collaboration between the architect and structural engineer achieved the gradually reducing brace dimensions up the height of the structure. Office building, Wellington.

compression, remembering that seismic forces change direction rapidly, then their cross-sectional dimensions must be sufficient to prevent buckling. The chords or end columns of braced frames also fall into this category for as well as resisting overturning-induced compression they also support gravity forces.

Braced frames, like shear walls, require strong foundations. But since the gravity forces acting on braced frames with their lighter construction are usually less than those of shear walls additional foundation stability might be required. Especially for slender one or two-bay frames, tension piles may be necessary to prevent over-turning. The structural engineer determines the type, diameter and length of a tension pile depending on the tension to be resisted and the ground conditions.

Materials and heights

The two most common materials of braced frames are wood and steel. Wood is appropriate for low-rise and light-weight construction. Although no structural reasons preclude its use as superstructure bracing, it is most often observed providing bracing to foundations (Fig. 5.28). Wood members are capable of resisting both tension and compression but large cross-sections are necessary to avoid buckling of long diagonals. Achieving the necessary connection strengths with bolts or other fastening devices is another challenge facing designers of wood bracing.

Steel tension and compression and eccentrically braced frames are suitable for most building heights. However, tension-only bracing with its inferior post-elastic behaviour as noted below, is normally restricted to low-rise construction.

Most braced frames are constructed from steel. The beams of eccentrically braced frames must be steel in order to provide the necessary ductility. Reinforced concrete or precast concrete columns and struts are sometimes used in eccentrically braced frames as well as for compression and tension braced systems. Although the member sizes of concrete braces are far less slender than those of steel they can be aesthetically pleasing (Fig. 5.29). Reinforced concrete braced frames are rather difficult to detail and construct. A large number of differently orientated bars intersect at the joints. Each bar also has to be adequately anchored with enough space around it for the concrete to be placed and compacted.

Braced frame ductility

Each of the three main types of braced frames has a different ductile mechanism as summarized in Table 5.4.

▼ **5.4** Braced frame types and comments on their ductility

Frame type	Material	Comments
Tension-only	Steel	Capacity Designed diagonal tension members should yield before any connections fail or the compression chord buckles. Compression diagonals buckle due to their slenderness while tension yielding members are permanently elongated. During an earthquake as both braces lengthen incrementally over successive seismic pulses, the structure becomes floppy and possibly unstable (Fig. 5.30). Tension-only bracing is best suited to low-rise construction.
	Wood	The strength of tension members is usually limited by connection strengths. Wood braces with conventional steel connections lack ductility and need to be designed to full elastic code forces.
Compression and tension	Steel	In a Capacity Designed system connection details are stronger than the tensile strength of a brace. Compression braces absorb very little earthquake energy. Ductility improves if approximately the same number of braces in tension and compression are in each line of framing.[6] Ductile fuses can also be inserted (Fig. 5.31). A fuse must be long enough to absorb inelastic axial deformations without tensile fracture yet short enough to compress plastically without buckling. Bucking-restrained braces that possess the characteristics of ductile fuses are discussed in Chapter 14.
	Wood	No significant ductility can be expected.
Eccentrically braced	Steel	In this highly ductile system the structural fuse is the short length of beam between inclined struts, or struts and adjacent columns. All connections and members are designed stronger than the fuse region to ensure it alone absorbs the earthquake energy. Closely-spaced vertical stiffeners along the fuse prevent the beam web from buckling.

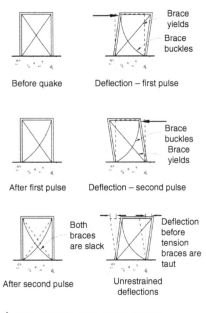

▲ **5.30** Under cyclic loading, yielding of tension-only braces result in a floppy structure.

▲ **5.31** Structural fuses in tension and compression bracing. Educational Institution, Wellington.

Deflected shapes (a) (b)

Shear force diagrams (c) (d)

Bending moment diagrams (e) (f)

Inertia force

Gravity force

▲ **5.32** Comparison between a horizontally loaded moment frame and a frame supporting gravity forces.

MOMENT FRAMES

Introduction

'Moment frame' is used in this book as a shortened version of 'moment-resisting frame'. While terminology varies between countries and some structural engineers simply prefer 'frame', the term 'moment frame' is intended to remind readers of the primary method moment frames transfer force – through bending. The three essential characteristics of a moment frame are: firstly, columns deep enough to resist significant bending moments; secondly, beams and columns with similar depths; and finally, rigid connections between columns and beams. Beams and columns of a moment frame are subject to relatively large bending moments as it drifts or deflects sideways while resisting seismic forces (Fig. 5.32).

Wherever bending occurs in a frame it is accompanied by shear force. The two are inseparable. Unless rigorously designed for seismic forces according to the Capacity Design approach columns, beams and beam-column joints fail in shear before bending (Fig. 5.33). Columns

▲ **5.33** Column shear failure in a moment frame. 1994 Northridge earthquake, California. (Reproduced with permission from A.B. King)

are especially vulnerable. They are subject to very high shear forces compared to columns of gravity resisting frames (Fig. 5.32 (c) and (d)). The end columns of seismic resisting frames can also experience significant tension and compression forces. This is how most of the overturning moment on a frame is resisted. The narrower and higher a frame the greater these axial forces become.

While Fig. 5.32 shows the effects of seismic and gravity forces separately, moment frames resist them simultaneously. Even where moment frames by virtue of their orientation with respect to gravity-resisting structure support no floor loads they still support their self-weight. So moment frames are always designed for the *combined* effects of seismic and gravity forces.

Like a shear wall a moment frame is effective in the direction of its length only. A one-way frame cannot resist forces at right angles to its length since there are no beams framing into the columns in that direction. Columns need to be orientated correctly with respect to the frame length to utilize their maximum strength and stiffness (Fig. 5.34). Where beams frame into columns from two usually orthogonal directions the columns become members of two frames (see Fig. 5.2(c)) and require additional strength.

Beam — Column   Beam — Column

Correct layout: Columns strong in direction of frame.
Frame effective for loading in this direction ➡

Poor layout: Columns weak and flexible when bending
about their minor axis

Reinforced Concrete   Steel

Roof slab ————   ———— Beam

▷A   ——— Beam-column
                                          joint
▷A

—— Column

—— Column
   depth

Elevation of typical reinforced concrete frame

Beam
depth

Section A-A

▲ **5.34** Column orientation in moment frames and frame elevation.

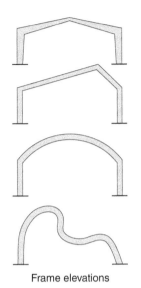

Frame elevations

▲ 5.35 Possible single-storey moment frame forms. Almost any shape is feasible structurally.

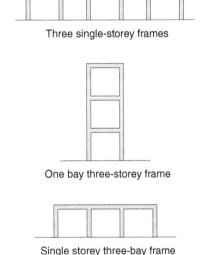

Three single-storey frames

One bay three-storey frame

Single storey three-bay frame

▲ 5.36 Single-bay frames as modules of multi-storey and multi-bay frames.

Moment frames can take a myriad of forms. So far one and two-bay rectilinear frames have been presented. Multi-bay frames are common and so are frames whose geometry breaks free from orthogonality to include pitched rafters or even embrace the curve. But provided certain structural requirements are met almost any scale or shape of moment frame is possible (Fig. 5.35). As illustrated in Fig. 5.36 three single-storey single-bay frames can be stacked vertically to form a three-storey one-bay frame, or joined sideways to form a single-storey three-bay frame. Architects in collaboration with structural engineers choose the layout of frames, numbers of bays, and bay width or distance between columns to suit architectural planning requirements. The chosen structural configuration must ensure sufficient overall strength and stiffness against seismic forces and hopefully enhance the realization of the architectural design concept. An architect should consider a moment frame, a shear wall, or even a braced frame for that matter, not only as a device to resist seismic forces but also as an opportunity to enhance the clarity with which an architectural idea or concept is realized.

It should come as no surprise that moment frames are such a popular method of providing seismic resistance. Moment frames meet the aspirations of contemporary building owners and inhabitants for minimum disruption to spatial planning, minimum structural foot prints and maximum opportunities for natural light and views. However, after having informed readers of the geometrical and configurational freedom associated with moment frames some necessary structural requirements must be mentioned.

Structural requirements

Relatively deep columns are the first prerequisite for a seismic moment frame. The column depth measured in the direction that the frame acts, and to a lesser extent the column width if it is rectangular in cross-section, must provide sufficient stiffness, bending and shear strength. To ensure ductile strong column – weak beam frames (to be discussed shortly) column depths are usually equal to or greater than those of the beams. Since reinforced concrete moment frames require special reinforcement detailing any column cross-section should be larger than 230 mm wide by 400 mm deep and even then such a small member might prove structurally inadequate for a building more than one-storey high.

Where columns are part of a two-way framing system they are normally square or circular in cross-section in order to possess sufficient strength in both directions. For a one-way frame cost-effective

▲ 5.37 Two suspended waffle slabs have collapsed. Retail store. 1994 Northridge earthquake, California.
(Bertero, V.V., Courtesy of the National Information Service for Earthquake Engineering, EERC, University of California, Berkeley).

columns are usually rectangular, contributing strength in one direction only. Columns need to be continuous from the foundations to the top of the moment frame at roof level.

A moment frame beam is approximately the same depth and width as the column it frames into to facilitate the direct transfer of bending moments between the two elements. The beam is also considerably deeper than a typical floor slab. The depths of moment frame beams are rarely less than their span/12 whereas slab depths typically range between span/25 to span/30. The beams of a moment frame can not take the form of slabs. Even waffle slabs have exhibited very poor seismic performance. During numerous earthquakes they have demonstrated their unsuitability to resist cyclic seismic forces sometimes by pancaking (Fig. 5.37). Slabs without beams are too flexible and weak to provide adequate lateral resistance even though they are perfectly adequate for gravity forces.

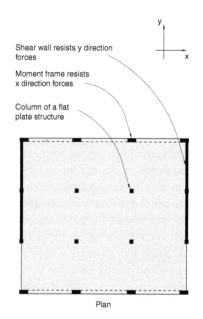

▲ 5.38 Plan of a flat plate gravity-resisting structure provided with seismic resistance by a pair of shear walls and moment frames.

Where either a flat plate or flat slab flooring system carries gravity forces a separate and recognized lateral force resisting system needs to be provided in each orthogonal direction (Fig. 5.38).

Ideally, moment frame beams should be continuous and form a straight line in plan. Horizontal off-sets along a beam line are to be avoided either along a beam or at columns. Beams, slightly curved in plan, may be possible structurally but straight beams avoid undesirable secondary effects. Beam centerlines should coincide with those of columns although codes do allow small offsets but not nearly enough for a beam to be attached to a column face.

Another defining feature of a moment frame is its rigid joints. As shown in Fig. 5.39, a moment frame requires at least one rigid joint. The more rigid joints the more evenly bending moments and shear forces are distributed around structural members and the greater the horizontal rigidity. Member sizes are also kept to a minimum. In most moment frames beam-column joints are designed and constructed to be rigid. One significant exception occurs in the case of single-storey frames. Columns are commonly pinned at their bases to facilitate

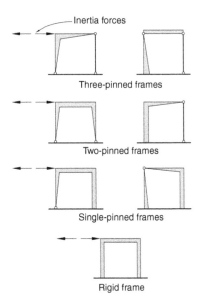

Inertia forces

Three-pinned frames

Two-pinned frames

Single-pinned frames

Rigid frame

▲ **5.39** Moment frame forms with different numbers of pins. Members are detailed to partially express their bending moment diagrams under seismic forces.

construction and reduce the need for enlarged foundation pads to withstand column bending moments. The penalty for this approach is that the frame is more flexible than if it were fully rigid so larger members are required.

Frames that are regular in elevation and plan display the best seismic performance. Regular frames comprise those with approximately equal bay widths and where all columns are oriented in the correct direction. Bay widths of multi-bay frames may be varied but the best configuration for seismic resistance is where beam spans are identical. Experience suggests an optimum distance between column centre-lines of between 5 and 8 m. Once a span exceeds 8 m deeper beams can require increased inter-storey heights.

Materials and heights of frames

Moment frames are fabricated from wood, reinforced concrete and steel. Glue-laminated and laminated veneer lumber (LVL) wood frames are reasonably popular in countries well endowed with forests. The main challenge facing designers is how to achieve rigid beam-column joints. Some jointing techniques are shown in Fig. 5.40.[7] Their complexity explains why the most practical rigid joint in sawn wood construction is formed with a diagonal brace. Wood moment frames are normally restricted to low-rise light-weight buildings (Fig. 5.41).

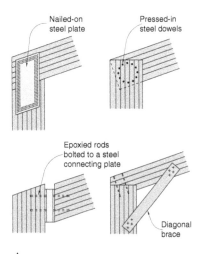

Nailed-on steel plate

Pressed-in steel dowels

Epoxied rods bolted to a steel connecting plate

Diagonal brace

▲ **5.40** Methods of forming rigid joints in (glue-laminated) wood moment frames.

▲ **5.41** An elegant wood moment frame. Commercial building, Austria.

Steel frames are suitable for light-industrial buildings through to high-rise construction (Fig. 5.42). Where reinforced concrete is the dominant structural material concrete moment frames can provide seismic resistance for buildings many storeys high (Fig. 5.43). Once over a

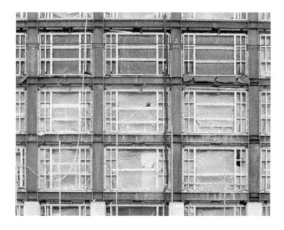

▲ **5.42** A perimeter steel moment frame. Apartment building, Wellington.

▲ **5.43** Reinforced concrete moment frames. Office building, San Francisco.

height of around 20 storeys they may require supplementing with shear walls or transformed into a different structural system such as a bundled-tube structure to restrict wind-induced drift.

Ductility

The two basic requirements of a ductile moment frame are: firstly, a ductile configuration and, secondly, ductile members. Application of the Capacity Design approach ensures both are satisfied as explained below.

Observation of seismic damage to frames shows time and again how poor frame configuration leads to concentration of damage in too few members such as ground floor columns that are incapable of absorbing the earthquake energy. Where damage occurs in columns they may be unable to continue to carry their gravity forces in which case collapse is inevitable.

Studies show that moment frames exhibit two failure mechanisms under seismic force overload. Firstly, plastic hinges or structural fuses can form at the top and bottom of the columns of just one storey, usually at ground floor level. In this case the earthquake energy is

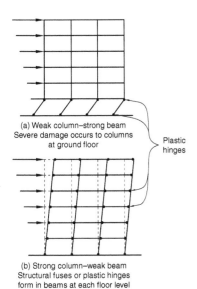

(a) Weak column–strong beam
Severe damage occurs to columns
at ground floor

Plastic
hinges

(b) Strong column–weak beam
Structural fuses or plastic hinges
form in beams at each floor level

▲ 5.44 Two potential overload
mechanisms of moment frames. Only the
strong column–weak beam configuration
is ductile.

▲ 5.45 A collapsed building with weak
columns and strong beams. Ironically
the architect enlarged the columns with
non-structural masonry. Mexico City, 1985
Mexico earthquake.
(Reproduced with permission from R.B. Shephard).

absorbed in just several locations and the most important load-bearing members of all, the columns, are badly damaged (Fig. 5.44(a)). The second, and ideal situation, is shown in Fig. 5.44(b). Plastic hinges form at the ends of many beams. This means less energy absorbed and less damage per hinge. Earthquake energy is now dissipated far more uniformly throughout the entire structure rather than concentrated in one floor. When a beam end forms a plastic hinge the beam can still support its gravity forces even though it is damaged. This is a far better situation than column hinging because beams, without significant compression forces within, are inherently more ductile than columns, and their damage does not jeopardize the safety of the entire building. The only hinges permitted in columns are structurally unavoidable and they form just above the foundations. Those hinge regions are specially detailed in the case of reinforced concrete construction with very closely-spaced ties to confine the concrete.

The scenario of Fig. 5.44(b) is an example of the application of Capacity Design. Beams are chosen as sacrificial members to dissipate the earthquake energy long before the columns are damaged. All other elements are designed to be stronger than the beam hinges. Columns are designed stronger than the beams and foundations stronger than the columns since foundation damage is difficult to detect and repair. Fig. 5.45 shows a collapsed building with columns smaller and weaker than the beams. As explained previously, damage that concentrates in the columns of a ground floor or any other single storey of a weak column–strong beam structure must be strenuously avoided.

As noted previously, a serious implication for architects adopting a ductile moment frame configuration is that the depth of columns must be approximately equal to or greater than the depth of beams. Reinforced concrete columns can be *slightly* smaller than beams provided that the columns contain more and/or stronger reinforcing steel.

Structural redundancy, a desirable configuration characteristic also increases ductility. More beam and column hinges mean less damage to each. Also, if one member fails prematurely, perhaps due to a construction defect, then the forces it was designed to resist can be shared by other intact members.

The terms 'plastic hinge' or 'structural fuse' describe the ductile energy-absorbing damage incurred where longitudinal reinforcing steel or a steel section in a moment frame yields plastically. Some good and poor examples of these hinges after strong quakes and full-scale laboratory tests are illustrated. Fig. 3.15 shows a damaged reinforced concrete column. Due to the large horizontal deflections it has undergone its cover concrete in the fuse region has spalled off but closely-spaced spiral ties successfully confine the core concrete and prevent shear failure. The column has performed in a ductile mode displaying the type of damage expected during a design-level earthquake.

Fig. 5.46 illustrates poor performance of a beam plastic hinge. The beam longitudinal steel has yielded in tension due to gravity and seismic bending moments, but because the beam ties are spaced too far apart large pieces of concrete have dropped from the core of the beam. The beam bars then buckled during reversed load cycles further weakening the beam and leaving the building in considerable danger should it be struck by a large after-shock.

A full-scale interior beam-column assemblage tested in a laboratory has been subject to cyclic loading to simulate the actions of a strong earthquake. Severe beam damage is localized in the plastic hinge region where the beams join the column (Fig. 5.47). Because the column and beam-column joint are stronger than the beam, damage is intentionally localized at the end of the beam. The beam longitudinal steel has been stretched plastically in tension and compression during the load cycles but closely-spaced ties have confined it well and prevented it from buckling. Similar damage is expected in beam plastic hinges of real buildings after a severe earthquake. This is what ductile behaviour looks like. Although structural damage has occurred the beam is almost as strong as it was before the quake.

How do designers achieve ductile steel moment frames? If not designed and detailed according to Capacity Design principles even steel frames

▲ **5.46** Poor example of a beam plastic hinge. The column is to the left. Mexico City, 1985 Mexico earthquake.
(Reproduced with permission from R.B. Shephard).

▲ **5.47** Full-scale laboratory test of a beam-column assemblage. A plastic hinge has developed after loading equivalent to a severe quake. Canterbury University.

are brittle. Columns must be stronger than beams. Hundreds of steel moment frames developed serious cracks where beams connected to columns during the 1994 Loma Prieta, California earthquake. Now more ductile details have been developed, including one where a potential plastic hinge is intentionally formed by locally weakening beam flanges (Fig. 5.48). Having created a fuse region, all other connections including welds are designed to be stronger so as not to suffer damage.

A final cautionary note: although designers may intend that moment frames be ductile, in practice ductility is difficult to achieve. Very high structural design and construction standards are necessary. Moment frame performance is sensitive to small design and detailing errors that can have grave consequences. If there is doubt about quality assurance standards consider using shear walls instead of frames to resist seismic forces.

▲ **5.48** Intentional weakening of a moment frame beam to form a structural fuse. The fuse region is weaker than any other structural member or connection. The wood flooring is unusual, but appropriate for this hotel built above an existing concrete parking building. Wellington.

MIXED SYSTEMS

Just as an artist mixing different coloured paints avoids a selection with a murky appearance, so an architect is careful about mixing structural systems to resist seismic forces. (The term 'mixed systems' applies where two or more different structural systems act together in one direction.) An example of mixed systems is where shear walls and moment frames act in parallel in the x direction in Fig. 5.49.

The problem with mixed systems, like the murkiness of an artist's colour, is a lack of clarity. With more than one system designers find it difficult to comprehend the force paths. Only through sophisticated engineering analysis can the combined structural behaviour of mixed systems be understood. The reason is that different systems are inherently structurally incompatible. With reference to Fig. 5.49, the wall is much stiffer and stronger than the moment frame. Even though a designer might intend the frame to contribute significantly to seismic resistance, because of its relative flexibility its effect is negligible. The wall effectively resists all the horizontal forces. Once the walls suffer damage the frames will pick up some force, but are they strong enough? A mixture of structural systems to suit architectural planning or some other imperative leads to a structural combination that is unintelligible to all except a structural engineer with a powerful computer – the antithesis of a structure with simple and clear force paths. Generally, architects should avoid mixed systems, except as outlined below.

▲ **5.49** An example of a mixed system comprising shear wall and moment frame.

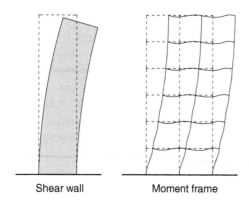

Shear wall   Moment frame

▲ **5.50**   The different deflected shapes of a shear wall and a moment frame resisting seismic forces.

A successful mix of structural systems occurs in high-rise buildings where walls and moment frames work together to resist horizontal forces. The inter-storey deflections or drifts of a moment frame are greater near its base than further up its height. The converse situation occurs with a shear wall (Fig. 5.50). Where acting in parallel with a wall, a frame resists most of the upper forces including inertia forces from the wall itself. As the forces travel towards the base of the combined structure they gradually move via floor diaphragms into the wall. Near the base of the building the wall resists most of the force. In the high-rise example of Fig. 5.51, each system compliments the other.

Another reason for mixing systems is to increase the *redundancy* of seismic resisting structure. The concept of internal forces finding alternative paths to the foundations in the event of one or more primary structural members failing is attractive – especially in the light of the widespread failure of steel moment frame joints during the 1995 Northridge earthquake as mentioned earlier. Elsesser suggests:

'*If carefully selected, multiple systems can each serve a purpose; one to add damping and to limit deflection or drift, the other to provide strength. Multiple systems also serve to protect the entire structure by allowing failure of some elements without endangering the total building.*'[8]

Several of the mixed systems he recommends incorporate structural devices, like dampers which are discussed in Chapter 14.

▲ **5.51**   A mixed structural system. A central shear wall core with a perimeter moment frame under construction. Office building, Wellington.

Some codes reward redundancy by allowing lower design forces. However, while increasing redundancy can be as straight forward as increasing the number of shear walls acting in one direction it can also be achieved by mixing structural systems. But due to the potential incompatibility problems outlined above and the need for a decision on an acceptable degree of redundancy to be informed by sound engineering judgment, mixed systems should first be discussed between an architect and an experienced structural engineer before being adopted.

Another form of mixed systems can occur where two or even three structural systems are placed above each other in a multi-storey building. For example, shear walls in the bottom few storeys might support and be replaced by moment frames that rise to roof level. Such changes in structural systems up the height of a building can introduce interesting architectural opportunities but they require especially careful structural design. The potential danger is the creation of one or more 'soft storeys' (refer to Chapter 9). The structural engineer must therefore rigorously apply the Capacity Design approach to achieve an overall ductile building. This is conceptually realized by adopting the principle of never allowing a weaker or more flexible structural system to be beneath one that is stronger or stiffer.

REFERENCES

1 Fintel, M. (1995). Performance of buildings with shear walls in earthquakes of the last thirty years. *PCI Journal*, **40**:3, 62–80.

2 Paulay, T. and Priestly, M.J.N. (1992). *Seismic Design of Reinforced Concrete and Masonry Buildings*, John Wiley & Sons, p. 402.

3 Brzev, S. (2007). *Earthquake Resistant Confined Masonry Construction*. National Information Center of Earthquake Engineering, Indian Institute of Technology, Kanpur, India.

4 Ghaidan, U. (2002). *Earthquake-resistant Masonry Buildings: Basic guidelines for designing schools in Iran*. Division of Educational Policies and Strategies Support to Countries in Crisis and Reconstruction, UNESCO.

5 Beca Carter Hollings and Ferner (1991). Indonesian Earthquake Study. Volume 6: Manual for the design of normal reinforced concrete and reinforced masonry structures. Unpublished.

6 Malley, J.O. (2005). The 2005 AISC Seismic provisions for structural steel buildings: an overview of the provisions. *STRUCTURE magazine*, **November**, 2–37.

7 Charleson, A.W. and Patience, D.B. (1993). Review of current structural timber jointing methods. *Journal of the New Zealand Timber Design Society*, **2**:3, 5–12.

8 Elsesser, E. (2006). New ideas for structural configuration. Proceedings of the 8th U.S. National Conference on Earthquake Engineering, April 18–22, 2006, San Francisco, California, USA, Paper No. 2182, pp. 9.

6

SEISMIC DESIGN AND ARCHITECTURE

INTRODUCTION

This chapter addresses how the previously discussed horizontal and vertical seismic resisting systems are brought together in the making of architecture. While the integration of structure with architecture in general is a challenge that architects face, the task of integrating seismic resisting structure is even more formidable due to the relatively large structural footprints required to resist seismic forces; at least, in regions of medium to high seismicity.

In order to appreciate the amount of seismic resisting structure that is sometimes required, consider a four- and an eight-storey office building in Wellington, New Zealand, founded on soft soils. The building materials are heavy; precast concrete slabs overlaid with topping concrete and with some concrete masonry interior and exterior walls. Figure 6.1 shows the structural footprints for the minimum vertical structure to support firstly, gravity forces, and then secondly, seismic forces.[1] For both buildings, the cross-sectional area of shear walls at ground floor level is approximately four times that of the gravity-only columns. If perimeter moment frames resist seismic forces in both directions their structural footprint to is up to three times larger than the gravity-only structural footprints. When the perimeter gravity-only columns are increased in size so seismic forces are resisted by two moment frames in each direction, then column dimensions increase by a factor of approximately between two and three. If wind force is the only horizontal force the shear walls are designed for, their cross-sectional areas reduce to less than one half the area required for seismic resisting shear walls.

The above example, set in an area of high seismicity and on soft soil – both of which lead to large design forces – illustrates how seismic resisting structure can significantly impact upon the structural footprint of a building. Seismic structure may profoundly affect an architect's

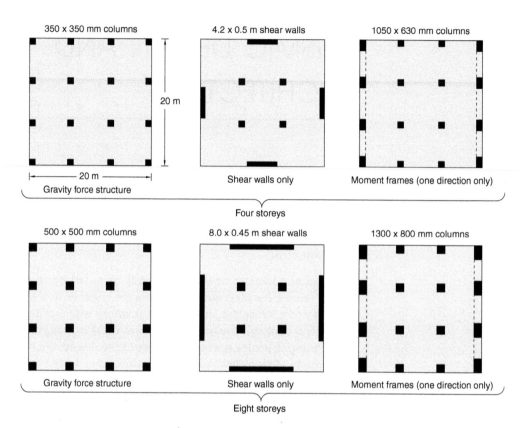

350 x 350 mm columns

20 m

20 m

Gravity force structure

4.2 x 0.5 m shear walls

Shear walls only

1050 x 630 mm columns

Moment frames (one direction only)

Four storeys

500 x 500 mm columns

Gravity force structure

8.0 x 0.45 m shear walls

Shear walls only

1300 x 800 mm columns

Moment frames (one direction only)

Eight storeys

▲ **6.1** Ground floor plans of a four and an eight-storey building showing the structural footprints first for gravity forces only, and then for seismic forces where resisted by shear walls and moment frames (drawn to scale).

1300 square columns at corners

1300 x 800 columns

▲ **6.2** The eight-storey moment frame building of Fig. 6.1 illustrating seismic resisting structure only. Although well-configured for seismic resistance it is architecturally bland.

ability to resolve both the design program and design concept satisfactorily. This chapter considers how such possibly dominant vertical structure is integrated with the architectural design and with other structural and non-structural aspects of a building.

INTEGRATING SEISMIC RESISTING STRUCTURE AND ARCHITECTURE

Readers may wonder if architecture in seismic zones might be bland and boring given its potential to be dominated by structural requirements (Fig. 6.2). As illustrated above, the size of seismic resisting structure can be large when compared to that required to withstand gravity and wind forces. Also, readers will remember recommendations from the previous chapter, such as wherever possible seismic resisting structures are symmetrical and regular and that moment frame column dimensions are larger than those of the beams. Surely the antithesis of exciting architecture!

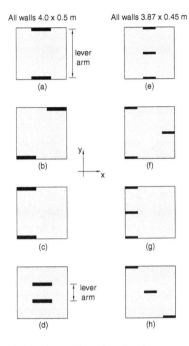

All walls 4.0 x 0.5 m   All walls 3.87 x 0.45 m

▲ 6.3 Ground floor plans showing different shear wall layout options that provide symmetrical resistance for x direction seismic forces (Y direction and gravity structure not drawn).

The intention of this section is to explain that, although special requirements of seismic structure must be met and might constrain some architectural intentions, these requirements may not be nearly as severe as first thought. Any negative influences may often be avoided by timely and creative structural configuration.

For example, consider the recommendation of structural symmetry. Fig. 6.3 shows the ground floor plan of the four-storey building discussed in Fig. 6.1. Two or three shear walls provide resistance against x direction forces. Even though each structure is perfectly symmetrical with respect to those forces some shear wall layouts do not appear symmetrical especially when considering forces acting in the y direction. This is because shear walls resist forces only parallel to their lengths. Of the four options suggested for a two-wall layout, all are equally effective although the lesser lever arm of option (d) reduces resistance against any in-plan torsion that may occur. All four three-wall options are equally satisfactory from both the perspective of x direction forces and torsion.

A similarly diverse structural layout is possible if, for example, the shear walls are replaced by one-bay or multi-bay moment frames. There are so many possible layouts, each of which can contribute in some way to a more expressive and compelling architecture. In Fig. 6.4, one-way frames in the x and y directions might both form and articulate circulation or assist in architectural form making. Figure 6.5 presents one possible layout for structure that is also symmetrical in both orthogonal directions. Over half the length of the perimeter walls

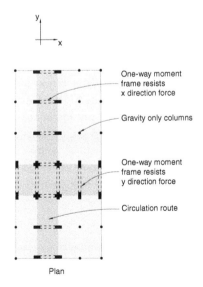

One-way moment frame resists x direction force

Gravity only columns

One-way moment frame resists y direction force

Circulation route

Plan

▲ 6.4 The creation of circulation routes using one-way frames whose structurally symmetrical layout does not cause torsion.

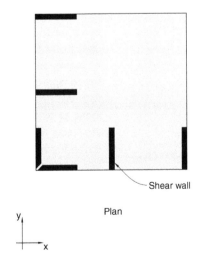

Shear wall

Plan

▲ 6.5 Ground floor plan showing seismic resisting structure placed symmetrically for x and y directions (gravity structure not shown).

as well as the floor plan is free of seismic resisting structure and hence suspended floors require minimally dimensioned gravity-only columns for support in those areas.

Architectural concept and planning

Two prerequisites for successful integration of structure with the architectural design concept and planning, are architects' attitude towards structure and a sense of timing. With a positive attitude which embraces the possibility of seismic structure enriching a design, an architect is more likely to engage with structural requirements soon after commencing the design process rather than procrastinating. The ideal situation is that, shortly after coming to terms with the program, an architect will ascertain the seismic requirements. For most designs, early discussions with a structural engineer will clarify the important issues. Then, armed with that knowledge, the architect can explore the opportunities and constraints of the seismic structure. While design ideas are still fluid, the question: 'How might this necessary structure contribute positively to the design?' can be posed. If this positive and timely approach is not followed, an unfortunate situation often arises where architectural design has to be reworked due to the unforeseen impact of seismic structure upon it.

This unsatisfactory approach to integrating architecture with seismic structure is summarized in Fig. 6.6. Model A, prevalent in contemporary architectural practice, describes a design method and sequence where structure is excluded from informing the architectural concept and form by virtue of its consideration late in the design process. Structure is an afterthought and its relatively late inclusion may be costly in terms of economy and how it might compromise architectural ideas. The preferred approach, Model B, tightly integrates structural considerations with both the realization of the architectural concept and the development of architectural form and planning. This design method enables architects to get the best aesthetic value from their structure as well as cost-effective and planning-friendly structural solutions.[2] Many possibilities of structure-enriching design can be explored.

Gravity resisting structure

As explained above, the architectural integration of seismic and gravity resisting structure and architect–structural engineer collaboration is best begun early in the design process. In the early days of seismic design when suspended floors were cast-in-place rather than utilizing

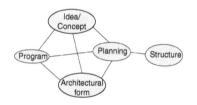

Model A

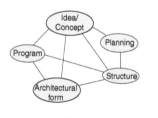

Model B
(Recommended)

▲ **6.6** Alternative models of the sequence and method of preliminary architectural design.

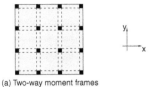

(a) Two-way moment frames

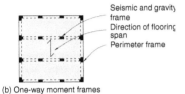

Seismic and gravity frame
Direction of flooring span
Perimeter frame

(b) One-way moment frames

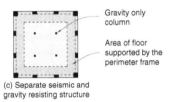

Gravity only column

Area of floor supported by the perimeter frame

(c) Separate seismic and gravity resisting structure

▲ **6.7** Floor plans showing different degrees of separation between seismic and gravity resisting structure.

precast concrete, seismic resistance of framed buildings was provided by two-way moment frames (Fig. 6.7(a)). Every column was a member of two orthogonal frames designed for gravity as well as seismic forces. Once designers appreciated the detailing and construction complexities and costs associated with ductile moment frames in the 1970s fewer structural elements were dedicated to resisting seismic forces. With reference to Fig. 6.7(b), if the floor structure spanning between the x direction beams is one-way there is no need for y direction beams to resist gravity forces. Since almost no floor loads are transferred to the two y direction perimeter frames they effectively can be designed to resist seismic forces only. Gravity and seismic resisting structure are therefore separated. In some buildings (Fig. 6.7(c)) the degree of separation between the two types of load-bearing structure is almost complete. Apart from a narrow perimeter ring of flooring carried by the perimeter frames interior columns support gravity forces while the perimeter frames resist all seismic forces. Not only do perimeter frames provide the best torsion resisting layout due to maximizing the lever-arm between them (see Fig. 5.5), but they possess two other advantages. Firstly, since seismic moment frame beams are relatively deep their perimeter location enables interstorey heights to be kept to a minimum. A potential clash between service ducts and the highly reinforced members is avoided. Secondly, by confining the large seismic resisting columns to the perimeter they are less disruptive to interior planning and can function as cladding elements, reducing façade costs.

The move away from two-way frames reduces structural redundancy. Now fewer elements provide seismic resistance. The search for economy of structure leads to the concentration of seismic structure rather than its more even distribution. Seismic and gravity structures are separated. The limit of this rationalization is reached with a seismic resistant structure consisting of two one-bay moment frames or two shear walls in each orthogonal direction (Fig. 6.8). Given the lack of redundancy, structural engineers need to be especially careful in the design, detailing and construction of the few critical structural elements and to design the slender gravity-only columns to accommodate horizontal seismic deflections without damage. Although these columns are not designed to resist seismic forces they do experience the same horizontal movements as the primary force resisting system.

Once the concept of separating seismic from gravity structure is accepted architects have more configuration options. No longer does a regular grid need to be superimposed upon a whole floor plan. Structural regularity in-plan, an indication of the degree of torsion and its resistance is assessed by moment frame, braced frame or shear

▲ **6.8** Seismic forces are resisted by two one-bay moment frames in each orthogonal direction. Shallow beams and slender columns support gravity forces. Office building, Wellington.

▲ **6.9** Slender gravity-only columns in the foreground with larger moment frame columns behind. Office building, Wellington.

wall location and not by its geometric symmetry as illustrated in Fig. 6.5. Gravity-only structural configuration can be treated more freely. It need not be 'gridlocked', and now that some columns do not resist seismic forces because of their slenderness and relative flexibility they are released from the strong column – weak beam rule. Columns can be slender and other materials explored (Fig. 6.9). Alternatively, columns can be enlarged but detailed as props – pin jointed top and bottom in each storey so as to not attract seismic forces to themselves and away from those moment frame columns designed to resist them.

As well as creating the opportunity to contrast the regularity and the large dimensions of seismic structure with more flexibly planned and slender gravity structure, separation of the two force resisting systems offers other design possibilities. For example, returning to Fig. 6.5 we see that the layout of structural walls is symmetrical about one diagonal. Moving along that imaginary axis away from the bottom left-hand corner one moves from opaqueness into openness or from possible intimate or private areas into a public or a more transparent realm. Peter Cook questions how structure is 'to be staccato, busy, cosy or symbolic of technicality?'[3] The structural diversity or hierarchy resulting from the separation of these structural systems invites architectural exploitation in search of answers.

Horizontal and vertical systems

A seismic resistant structure necessitates integration of horizontal and vertical structure. Together, these two systems resist inertia forces and transfer them along force paths predetermined by designers to the

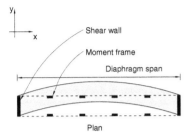

(a) Diaphragm deflecting under y direction inertia forces. Since its span is excessive for its depth it may be too weak or flexible

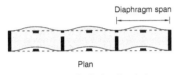

(b) More even distribution of vertical structure reduces the diaphragm span

▲ **6.10** A long diaphragm needs more frequent support from vertical structure.

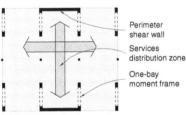

(a) One-way, one-bay frames allow distribution of under-floor services

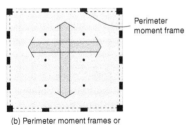

(b) Perimeter moment frames or shear walls allow maximum freedom for services distribution

▲ **6.11** Floor plans showing structural layouts to facilitate mechanical services distribution.

foundations. In most buildings, as explained in Chapter 4, suspended floors provide adequate diaphragm strength and stiffness without needing significant modification. But where the horizontal distance between vertical systems lengthens, the structural adequacy of a diaphragm warrants special attention (Fig. 6.10).

Where architects penetrate or perforate a diaphragm or vary its levels along the length of a building, these changes weaken it and reduce its ability to span horizontally. In these cases, the plan distribution of vertical structure needs to be carefully addressed. More information on how to deal with diaphragm irregularities is presented in Chapter 9, but in general terms, wherever the monolithic and planar character of a diaphragm is disturbed, architects need to provide more uniformly distributed or closely-spaced vertical structure.

Seismic structure and mechanical systems

In most building projects the architect, structural and mechanical engineer collaborate in order to integrate structure and mechanical systems. This challenge is heightened by the presence of seismic structure. Firstly, certain areas of structural elements such as structural fuses or potential plastic hinge regions, cannot be penetrated at all. Secondly, members like the beams and columns of moment frames are larger than normal. Moment frame beams may be so deep that there is insufficient depth between them and the intended finished ceiling height for any services to pass under. Due to highly congested reinforcing steel — particularly in beam-column joints — designers may not have any space to locate vertical pipe-work in columns of moment frames.

Just as a plea was made for early integration of structure and architecture earlier in this chapter, strategies for integrating structural and mechanical systems also need early resolution to avoid time and cost over-runs. In a highly serviced building, like a laboratory or a hospital, mechanical servicing issues have a huge impact on structural configuration. Internal moment frames, whose deep beams may disrupt intended service layouts, perhaps need to be confined to certain locations or substituted by more penetration-friendly structural systems like braced frames. Perhaps moment frames should be confined to the perimeter of the building (Fig. 6.11).

How much structure is needed?

When architects design gravity-laden structure at a preliminary design stage they can refer to plenty of design aids to help them size members

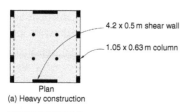

4.2 x 0.5 m shear wall

1.05 x 0.63 m column

Plan
(a) Heavy construction

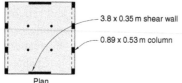

3.8 x 0.35 m shear wall

0.89 x 0.53 m column

Plan
(b) Medium construction

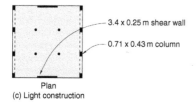

3.4 x 0.25 m shear wall

0.71 x 0.43 m column

Plan
(c) Light construction

▲ **6.12** Ground floor plans of a four-storey building showing how structural dimensions change for different construction weights (drawn to scale).

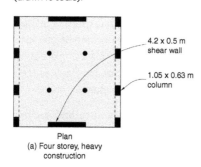

4.2 x 0.5 m shear wall

1.05 x 0.63 m column

Plan
(a) Four storey, heavy construction

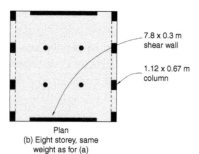

7.8 x 0.3 m shear wall

1.12 x 0.67 m column

Plan
(b) Eight storey, same weight as for (a)

▲ **6.13** Ground floor plans of a four and eight-storey building of the same weight (gravity structure not shown. Drawn to scale).

like beams, trusses and columns.[4] Unfortunately, similar information for the sizing of seismic structure is unavailable. The reason is simple – there are just too many variables that significantly affect structural dimensions. Yet architects do have some options for approximating member sizes before getting a structural engineer deeply involved. We now consider the primary factors that affect the amount of seismic structure that buildings require. In each case reported upon, below the structure is assumed to be fully ductile and therefore possesses the smallest possible footprint.

Building weight

The inertia force acting upon vertical structures like shear walls or moment frames is related to building mass or weight. If the four-storey building of Fig. 6.1 constructed from heavy materials is redesigned assuming medium and then light-weight construction its structural dimensions reduce significantly (Fig. 6.12). A medium-weight construction consists of steel framing supporting concrete flooring with a mixture of light-weight framed and heavy interior and exterior walls. Light-weight construction assumes wood flooring and walls.

Building height

If the four-storey building of Fig. 6.1 is increased in height to eight storeys, yet maintains the same total weight by using lighter building materials, then Fig. 6.13 shows that the shear walls need to almost double in length; yet the moment frame dimensions increase only marginally. As the building height increases, while keeping the weight constant, the natural period of vibration increases. This reduces the acceleration response of the building and the consequent design inertia force, as discussed in Chapter 2. However, the reduction in inertia force is insufficient to counteract the increased overturning moments due to the greater building height. More sensitive to overturning moment than the frames which are 20 m long, the relatively short walls require a greater increase in dimensions to cope with the additional building height. For both walls and frames horizontal deflection, rather than bending or shear strength, is the critical structural criterion determining final member sizes.

Seismic zone

The seismic zones of most earthquake-prone countries reflect varying levels of seismicity. In large countries like the USA and India, seismicity varies from very high to zero. In areas where the seismic hazard is low, designers can ignore seismic effects and wind forces become the dominant horizontal design force. But some countries' structural codes still wisely require minimum detailing standards – at least for reinforced concrete construction – to provide some ductility or resilience should an unexpected quake occur.

If the four-storey Wellington building of Figs. 6.1 and 6.12(a) is built in Auckland, which is located in the lowest New Zealand seismic zone, then the level of seismic design force reduces by 50 per cent. The new shear walls are now half their original area and the moment frame column dimensions reduce significantly.

Soil conditions

Chapter 2 explains how the type of soil beneath a building affects seismic performance. Compared to hard soil or rock, soft or flexible soils amplify bedrock accelerations. Buildings on soft soils require more structure than those built on rock. The amount of additional structure depends on the requirements of a country's seismic design code. If the building of Fig. 6.1 were to be relocated from a soft-soil site to a rock site, its seismic structural footprint approximately halves.

Numbers of structural elements

As a designer increases the number of elements, like shear walls, that resist seismic forces the total structural footprint increases. Figure 6.14 shows the trend when the number of shear walls for the Fig. 6.1 building is increased. The main reason for this increase is due to the fact that the structural stiffness of a member like a slender shear wall or a column is proportional to its depth cubed (depth^3). The horizontal stiffness of, say, a 2 m deep column is equivalent to the combined stiffness of eight 1 m deep columns, yet the 2 m deep column has only one quarter of their structural footprint.

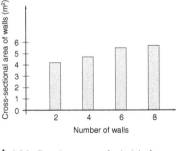

▲ **6.14** The plan area or footprint of seismic resisting structure increases with the number of shear walls. This graph applies for one direction of loading.

As the number of seismic resisting members reduces, so does structural redundancy. Where codes penalize a lack of redundancy by increasing design forces the results presented in Fig. 6.14 do not apply.

Towards an answer

Bearing in mind the five factors considered above it is not surprising there is no easy answer to the original question: 'How much structure does a building require?' There are just no simple rules-of-thumb. An accurate answer can be provided only after a structural engineer has performed numerous calculations, probably using computer software. It can be assumed that architects either lack the expertise to undertake such detailed design or do not want to. However, at a preliminary design stage, when approximate structural member sizes only are necessary, a structural engineer, or even an experienced architect who has developed a 'feel' for structural requirements, can make a reasonable 'guestimate'.

If engineering advice is unavailable then the options to determine approximate structural dimensions are limited. Precedents are helpful. Study the structures of nearby buildings – being careful to choose those most recent and, therefore, likely to have been designed to a modern seismic code. Make allowances as you see fit for differences in building weight and height.

The only other method of structural sizing at a preliminary design stage is to use specialist software. At present only one program, RESIST, is suited to the specific needs of architects; namely, simple non-numerical input and graphic output.[5]Figure 6.15 shows an example of one of its simple input screens. Originally developed for New Zealand schools of architecture and civil engineering by the author, RESIST has been customized for other countries such as India and Peru. Versions for China, the USA and Europe are under consideration. The advantage of a program like RESIST is that it empowers architectural students and architects to personally undertake preliminary designs of seismic resisting structure. Within a few minutes they can investigate the suitability of many different systems and configurations in order to choose a solution best integrated with their architectural aspirations.

SPECIAL STRUCTURES

So far in this book only three structural systems have been mentioned: shear walls, braced frames and moment frames. Although these are by far the most common seismic resisting systems, several architectural forms rely on other systems for seismic resistance. Examples include tension membranes and shell structures.

Tension membranes are usually so light-weight that inertia forces are much smaller than wind forces. If a tension membrane meets the structural requirements for gravity and wind force you can usually assume that it will perform satisfactorily during a quake.

Shell structures resist and transfer forces through tension and compression forces within the thickness of their shells. The strength of shells depends upon a geometry that facilitates axial force transfer and

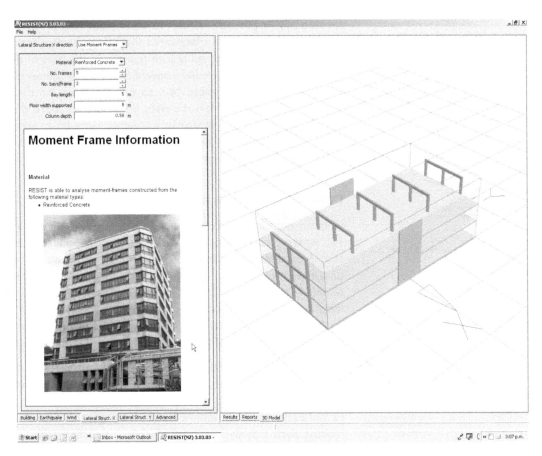

▲ **6.15** A screen print from the RESIST software. To the left the only input screen to design a reinforced concrete shear wall, and to the right, a model of the building. Analysis results, on another screen, are presented as bar charts.

minimizes bending moments. Shells are constructed as either continuous rigid curved surfaces, fully triangulated members as in a geodesic dome or as a lamella roof, comprising two opposed skewed grids of arch-like elements. Compared to more conventional structural forms, the gravity force paths of shells are complex and necessitate sophisticated computer analyses. The addition of seismic forces further compounds the difficulty of structural design.

There are certainly no rules-of-thumb for the preliminary design of these structures against seismic forces. Structural engineering advice is essential. However, it is possible for designers to draw analogies with more common structural systems. For example, the seismic resisting structure of a dome subject to horizontal force can be considered

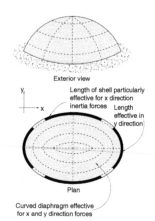

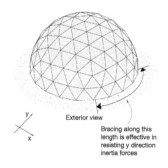

▲ **6.16**   A simplified approach to understanding the horizontal resistance of a rigid shell structure.

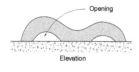

▲ **6.17**   A simplified approach to a triangulated shell structure.

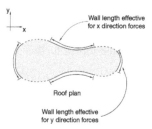

▲ **6.18**   Bracing lengths for an irregular shell structure.

simplistically in plan as two curved shear walls (Fig. 6.16). A similar approach is applicable to triangulated shells. Their surface areas can be considered as doubly-curved braced frames. At the base of the structure where the frames are like conventional vertical frames they need adequate numbers of braced bays to transfer forces to the foundations (Fig. 6.17). Away from the walls roof areas act more like curved diaphragms. Where a shell structure is rather irregular, an appreciation of its seismic resistance can be gained by identifying lengths of 'shear walls' at ground floor level (Fig. 6.18).

CONTEMPORARY ARCHITECTURE IN SEISMIC REGIONS

Without entering the territory of Earthquake Architecture (Chapter 17), where ideas and concepts expressive of seismic effects inspire architectural designs, a brief examination of contemporary architecture in seismically active regions is warranted. Is the architecture of seismic regions different from that of seismically quiescent areas? Do the rules and recommendations regarding regularity, symmetry and so forth necessitated by seismicity, exercise a stultifying influence upon architecture and cause it to become less interesting and exciting?

Regarding the first question the relevant literature remains silent. No studies have, for instance, compared the architecture of, say, two cities with very different levels of seismicity, like New York and Los Angeles, to find evidence of any seismic factor that might influence their architecture. Such a study would be complex. And how does one even begin to assess architectural interest and excitement? So many factors influence built form – the immediate and wider built environment, cultural, climatic, constructional and historic aspects, to name but a few.

One starting point to engage with these questions is to review publications featuring the contemporary architecture of cities or countries located in seismic zones, such as Los Angeles and Japan. A cursory scan through the glossy pages of typical journals suggests that contemporary practicing architects are by no means overly constrained by 'seismic rules'.[6,7] Compositional irregularity is common among architecturally significant low-rise Los Angeles buildings. Their architectural exuberance is partially explained by the fact that many are located on green-field rather than urban infill sites. Off-set and sloping walls and columns, unsymmetrical, splintered, fragmented and curved forms abound. One example is the work of architect Eric Owen Moss. Two of his buildings on adjacent sites exemplify irregularity and complexity of form (Fig. 6.19). In few projects one is aware of a dominant or ordering structural rationale, but it is noticeable that a predominance

▲ **6.19** The Stealth Building whose cross-sectional form morphs from triangular to square along its length. Los Angeles.

of light-weight construction reduces the potential architectural impact of the requirements of seismic structure. Irregularity also features in Japanese examples with their juxtaposition of geometric forms, long cantilevers and sculptural qualities. Generally, Japanese construction materials are heavier and, while there is greater visual evidence of seismic resisting structure, once again there is no sense of, nor mention of architectural creativity having been hindered by seismic design requirements.

Irrespective of these findings there is no avoiding the fact that buildings in moderate- to high-seismicity regions require increased horizontal strength and stiffness in two orthogonal plan directions, as well as torsional resistance. Additionally, the requirement for a hierarchy of member strengths, particularly in moment frames with their weak beams and strong columns, set seismic resistant structures apart from structures in non-seismic regions. Seismic design entails meeting certain minimum requirements specified by the structural codes of each country. So why are these requirements not having a visible impact upon the architecture?

The first reason is that often seismic resisting structure, just like gravity structure, is concealed. Consider the San Francisco Museum of Modern Art, designed by Mario Botta and completed in 1995 (Fig. 6.20). Internal steel braced frames ensure the structure complies with the requirements of Californian seismic codes. The frames are concealed between

▲ **6.20** San Francisco Museum of Modern Art.

▲ **6.21** Braced steel structure of the San Francisco Museum of Modern Art during construction.

▲ **6.22** Separation joint between panels of brick masonry filled with flexible sealant. San Francisco Museum of Modern Art.

wall linings within the interior of the building and placed behind exterior masonry walls (Fig. 6.21). Although Botta's characteristic deep slots and recesses might signify that the exterior masonry walls are not load-bearing it is still difficult to appreciate that the building is steel framed. The 'solid masonry' is no more than a skin of loosely attached masonry panels separated from each other and adjacent structural elements to avoid damage when the building sways in a quake (Fig. 6.22). Not only are the seismic resisting elements completely hidden from view but one presumes wrongly that the building structure, at least in part, consists of load-bearing masonry.

Even if primary structure is concealed, keen observers can still recognize subtle tell-tail signs of seismic design. But these are little more than clues and so are difficult to discern during a building visit let alone from photographic images. For example, we have just noted the separation details between cladding panels at the San Francisco Museum of Modern Art. Window mullions in another building might be wider than normal to accommodate seismic movement. A newer building is set back from its boundaries to allow for seismic drift, the gap screened by flexible flashings. A large complex might be separated into structurally independent blocks. These finer details of seismic design certainly exist but are not particularly evident.

Light-weight construction is another reason explaining the lack of visible application of seismic requirements. Light-weight walls, perhaps

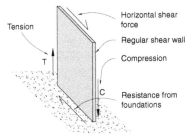

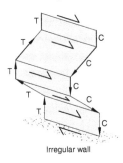

▲ 6.23 Comparison of the load path down a regular wall with that of an irregular wall at the Seattle Public Library. The additional structure required where forces change direction is not shown.

framed by wood or cold-formed steel studs, and wooden floors and roof attract little inertia force. Even if wind forces do not dominate horizontal loading, smaller forces mean less structure and make it easier and less expensive to design the complex force paths generally associated with exciting architecture.

When an architect breaks away from regularity and symmetry, force paths become more complex. More demands are made upon the elements of seismic systems. Expect increased numbers of transfer diaphragms, collectors and ties, and mixed systems, like moment frames working together with shear walls in the same orthogonal direction. Compare the force path through a geometrically irregular wall of the Seattle Public Library (opened in 2004), and designed by OMA, with that of a regular and continuous wall (Figs. 6.23 and 6.24). Additional structures to stabilize those areas where forces change direction along their path can be extensive and expensive. Complex force paths require sophisticated engineering design.

▲ 6.24 Seattle Public Library.
(Reproduced with permission from Maibritt Pedersen).

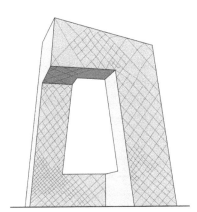

▲ 6.25 CCTV Headquarters, Beijing.

The new CCTV headquarters in Beijing provides one of the most extreme examples of structural rigour necessitated by complex force paths (Fig. 6.25). As one might expect given its scale, geometric complexity and its inherent instability, structural engineers undertook a huge number of exceedingly complex analyses to develop and optimize the structural system. Then they had to demonstrate through computer modelling satisfactory seismic performance for each of three increasingly severe earthquake scenarios.[8]

Undeterred by the intrinsic difficulty of answering the question posed at the beginning of this section regarding possible architectural limitations posed by seismic design requirements, a group of sixty fourth-year architectural students have undertaken what might be considered a pilot study. Half the students chose cities in the most seismically active areas of the world, while the other half selected cities with no significant seismicity. Every student then chose on the basis of class-decided criteria the best example of contemporary built architecture in that city and analysed any seismic configuration problems (these topics are covered in Chapters 8 and 9). When the students assessed the degree of architectural interest of each others' buildings without knowing the locations of the buildings, there was no discernable difference in the degree of architectural interest for both groups of buildings. Nor were there any significant differences in configuration irregularities between them.[9]

CASE-STUDY: THE VILLA SAVOYE

As an example of meeting the requirements for seismic resisting structure and the challenge of integrating it architecturally, consider one of the iconic buildings of the Twentieth Century: Le Corbusier's Villa Savoye. Completed in 1929, it is located in the seismically benign city of Paris (Figs 3.6 and 6.26). What if an identical building were to be built in a seismically active city like Tokyo or Istanbul? The gravity forces of the Villa are carried by reinforced concrete flat slabs including a few beams, and columns. In the original building horizontal wind forces are resisted by plastered unreinforced concrete masonry walls – the blocks laid after casting the surrounding reinforced concrete posts and beams. In both storeys the walls are reasonably well-distributed in plan, although at ground floor the centre of resistance lies towards the rear of the building, creating torsional eccentricity. Very few walls are continuous up the height of the building. Diaphragms at both levels are penetrated by the ramp and the roof diaphragm has lost most of its integrity due to cut-outs for the two terraces (Fig. 6.26(d)).

How to provide dependable seismic resistance? If the building were of wood construction it might be possible to treat each existing wall as a plywood shear wall. This could provide sufficient bracing in each orthogonal direction. Since ground and first floor walls are off-set, the first floor diaphragm would have to function as a transfer diaphragm. Additional floor structures – like transfer beams – would be required to resist the overturning-induced tensions and compressions from the chords of each of the first-floor walls and transfer those point forces to the ground floor columns and foundations. Since the Villa Savoye is of heavy construction, the preferred seismic resisting structure is one that is more rational with simple and direct force paths.

The design approach taken to provide seismic resistance is to assume all existing walls are non-structural. For a re-built Villa they would be constructed from reinforced masonry designed to withstand

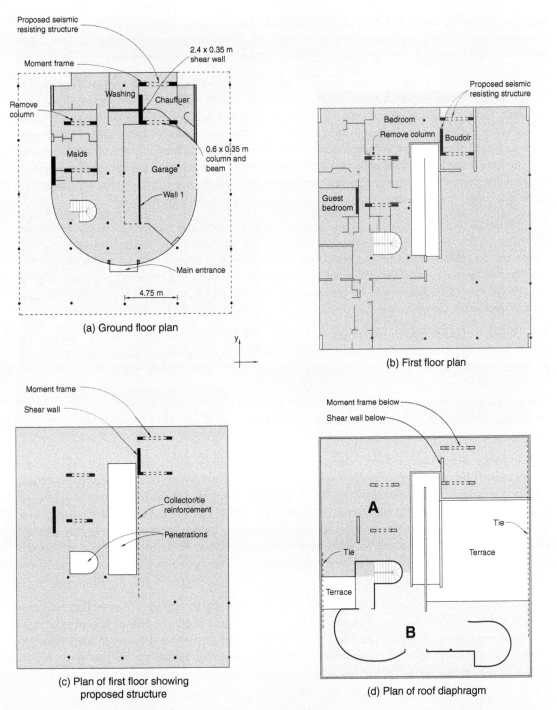

Proposed seismic
resisting structure

Moment frame

2.4 x 0.35 m
shear wall

Remove
column

Washing

Chauffuer

Maids

0.6 x 0.35 m
column and
beam

Garage

Wall 1

Main entrance

4.75 m

(a) Ground floor plan

Proposed seismic
resisting structure

Bedroom

Remove column

Boudoir

Guest
bedroom

(b) First floor plan

Moment frame

Shear wall

Collector/tie
reinforcement

Penetrations

(c) Plan of first floor showing
proposed structure

Moment frame below

Shear wall below

A

Tie

Tie

Terrace

Terrace

B

(d) Plan of roof diaphragm

▲ **6.26** Simplified floor and roof plans of the Villa Savoye showing the proposed seismic structure if rebuilt in a seismically active area.

out-of-plane inertia forces and structurally separated, as explained in Chapter 10. Due to their small diameter, the columns (or pilotis as they are usually called) are far too weak and flexible to function as members of moment frames assuming that there are beams, which in most cases there are not.

Now begins the difficult task of finding locations and space for the new structural systems. They need to be continuous from foundations to roof without drastically affecting existing spatial planning. One solution is to design two shear walls to resist y direction forces (Fig. 6.26(a)). From preliminary calculations performed by RESIST assuming a soft-soil location in Wellington, two ductile reinforced concrete walls 2.4 m long by 0.35 m thick are sufficient. If the bracing effect of existing Wall 1, which occurs only at the ground floor, is taken advantage of by making it structural the dimensions of the two new walls can be reduced. The rear wall passing through the chauffeur's space and the boudoir above has little impact on planning but the left-hand wall significantly reduces light into the maids' room.

It is impossible to use shear walls in the x direction without destroying interior spaces or compromising the architectural form. The chosen approach therefore is to use four one-bay moment frames. Of all options they best suit the existing planning. Unfortunately, due to their 600 mm column depth they are intrusive in some spaces like the boudoir and disrupt the maids' room. Its existing column which already disrupts that space would be removed but it is not nearly as inconvenient as the proposed column over three times its size.

The difficulty of inserting the proposed structure within existing subdivided space, as well as the rather unsatisfactorily forced and awkward outcome, clearly illustrates a point made previously. Preliminary spatial planning must be undertaken simultaneously with the development of rational seismic structure. If structural and planning requirements inform the development of each other, the outcome is a scheme where structure and interior spaces are harmoniously integrated. Otherwise, if these requirements are addressed independently the outcome as (in this example) leaves a lot to be desired.

The final step in the design of the relocated Villa Savoye is to assess diaphragm adequacy. Can the heavily penetrated diaphragms transfer inertia forces from all areas in plan to the proposed new shear walls and moment frames? For the first floor diaphragm, the answer is 'yes'. Both penetrations will not affect inertia forces being transferred into the shear walls. Since the right-hand shear wall is near the rear of the building, and shear forces cannot be effectively channelled into it where it is adjacent to the penetration, a collector is provided, as shown in Fig. 6.26(c). It consists of several continuous reinforcing bars anchored in the wall and embedded in the concrete slab beyond the length of the penetration.

The roof diaphragm is far more heavily penetrated (Fig. 6.26(d)). It is almost severed into two sections, A and B. A collector or tie member strong in both tension and compression is needed to connect both areas along the right-hand side. Then approximately half the inertia forces in the y direction from area B can be transferred to an area of diaphragm strongly connected to the rear shear wall. A similar tie is required along the left-hand side to transfer inertia forces from the left-hand side of area B into the shear wall on that side of the building.

Diaphragm performance in the *x* direction is more uncertain (Fig. 6.27). Note the considerable torsional eccentricity and moment in this direction. All the inertia forces from area B have to be transferred in shear through the 1.7 m wide strip between the smaller terrace and the circular stair penetration (Fig. 6.26(d)). Special calculations are required to ascertain if that highly stressed area can be made strong enough. Certainly many extra reinforcing bars are needed. Only then can the safe transfer of shear force from area B to the four moment frames be guaranteed. Another issue to be considered is the in-plan bending moment created about the weakened section when area B inertia forces act in the *x* direction (Fig. 6.27). Although some of that moment can be resisted by the two strips of diaphragm connecting areas A and B on either side of the stair they are probably too weak. The two *y* direction perimeter ties assist. The right-hand tie experiences compression and the left-hand tie tension for the loading direction shown. Both ties need to be checked to determine they can sustain the design compression forces without buckling.

In summary, it is possible to recreate the Villa Savoye in a seismically active area. But the proposed structural solution is far from elegant given the incompatibility of seismic resisting structure with existing interior planning. The case-study warns of the negative consequences that arise when architects procrastinate engaging with seismic design requirements.

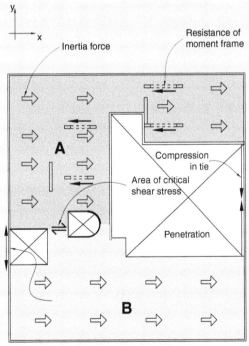

Plan of roof diaphragm

▲ **6.27** Inertia forces acting in the x direction; the resistance from the proposed structure and the axial forces in the two perimeter tie members.

REFERENCES AND NOTES

1 The amount of structure required to resist horizontal forces is determined using RESIST, a preliminary structural design computer program described near the end of the chapter. Ductile shear walls and frames are assumed and forces are in accordance with the New Zealand Loading Standard, NSZ 4203:1992.

2 For a detailed exploration of how structure can contribute architecturally other than by its load-bearing function see Charleson, A.W. (2005). *Structure as architecture: a sourcebook for architects and structural engineers*. Elsevier.

3 Cook, P. (1996). *Primer*. Academy Editions, p. 85.

4 The following reference is one of many that provides approximate structural member dimensions: Schodek, D.L. (2001). *Structures* (4th Edn). Prentice-Hall, Inc.

5 Editorial review (2004). 'RESIST' − structural modeling software. *Connector*, Vol. XIII, No.1, 12–13.

6 Chase, J. (2006). *LA 2000+ New architecture in Los Angeles*. Monacelli Press.

7 Meyhöfer, D. (ed.) (1993). *Contemporary Japanese architecture*. Benedikt Taschan.

8 Carroll, C. et al. (2005). CCTV Headquarters, Beijing, China: structural engineering design and approvals. *The Arup Journal*, **2**, 3–9.

9 Charleson, A.W. Comparison between contemporary architectural form in cities with high versus low seismicity. Awaiting publication in *Earthquake Spectra*.

7 FOUNDATIONS

INTRODUCTION

Particularly when presented with a green field site, and even before
the concept design is begun in earnest, an architect needs to be aware
of how ground conditions might impact site selection. An awareness
of ground hazards associated with earthquakes provides an opportunity at the concept stage to site a building so as to avoid the ground
hazards and foundation problems discussed later in this chapter. For
example, by moving a site back from steep hillsides prone to landslides
or from river banks likely to experience lateral spreading otherwise
expensive and time-consuming ground works may prove unnecessary.

Site selection requires a team approach. The architect and structural
engineer need input from a geotechnical engineer who assesses all
likely ground hazards, which not only may affect the building site itself
but also access to it by vehicles and services. For important building
projects like hospitals site selection requires a particularly thorough
approach.[1] After assessment and selection of a site, attention must
then be directed to how the building is to be founded.

Unless architects are designing standard single or double-storey dwellings, they rarely draw foundations on their plans. Although hidden, buried beneath the ground surface, foundations are essential. They interface
superstructure and ground, distributing highly concentrated forces from
concrete or steel members to the ground without overstressing it.
While foundations are the elements through which buildings are safely
'grounded', they are also the area through which potentially damaging ground shaking enters a building. Foundations transfer earthquake
energy into, and to a lesser degree, out of a building. During this process
they should remain undamaged and prevent damage to the underlying
supporting ground. In combination with the soil that encapsulates them,
foundations are designed to resist sliding and overturning. Figures 5.9

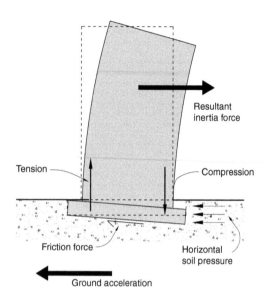

▲ 7.1 Forces acting on a foundation from the superstructure and the ground.

and 7.1 illustrate how overturning leads to increased compression forces at one end of the foundations and possibly tension at the other.

As discussed in Chapter 2, the soil conditions at a site affect the intensity of earthquake shaking. Soft soils amplify bedrock shaking and lengthen the period of vibration felt by a building. But in a far more sinister manner certain soils react poorly to shaking. Seismic vibrations change their engineering properties, and always for the worse as far as buildings founded upon them are concerned.

Involvement of structural and geotechnical engineers greatly reduces the risk of a quake disrupting the safe founding of a building. While structural engineers are competent to design foundations for low-rise buildings on geologically stable sites, for high-rise construction and poor soil conditions a geotechnical engineer should be consulted. Although there is seemingly nothing more basic than rock or dirt, where these natural materials support buildings they perform vitally important structural roles and their dynamic behaviour can be exceedingly complex.

Although (for other than low-rise domestic buildings) an architect does not usually get involved in the details of foundation design, as a member of and perhaps the leader of a design team, an architect needs to appreciate foundation investigation and design processes in order to communicate aspects of them to clients. If the lead designer, an architect recommends a program of foundation investigations to a client on the basis of engineering advice, and requests financial approval to proceed.

SEISMIC FOUNDATION PROBLEMS AND SOLUTIONS

Before discussing the topic of foundation investigations that precedes any foundation design, this section outlines common seismically-induced foundation problems that site investigations and the subsequent design are expected to identify and remedy. An appreciation of potential problems enables the scope of investigations to be fine-tuned and undertaken efficiently, and for appropriate foundation design to be undertaken.

Liquefaction

The term 'liquefaction' describes the phenomenon whereby perfectly stable soil underlying a flat site and possibly having provided a

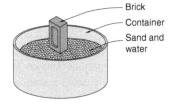

Brick
Container
Sand and
water

▲ **7.2** A simple demonstration of liquefaction.

▲ **7.3** A building after liquefaction. Dagupan City, 1990 Philippines earthquake.
(Reproduced with permission from David C. Hopkins).

solid founding stratum for hundreds of years suddenly liquefies during a quake. Fortunately, most soils are not in danger of such a rapid phase-change with its subsequent destructive effects on the built environment. The three main prerequisites for liquefaction are: a layer of relatively loose sand or silt, a water table high enough to submerge a layer of loose soil, and finally, an intensity of ground shaking sufficient to increase the water pressure between soil particles to cause the soil-water mixture to liquefy. Even if the ground shaking does not cause 'liquefaction', the increase in water pressure can lead to a reduction in the ability of foundation soils to withstand loads from buildings.

Liquefaction can be easily demonstrated by filling a bucket with dry or moist sand, and pouring in water. A brick placed upright in the sand tilts and sinks when the side of the bucket is tapped with enough force (Fig. 7.2). A sandy beach provides another opportunity to experience liquefaction. By rapidly vibrating your feet at the water's edge the sand–water mixture liquefies. The sand beneath your feet settles as it consolidates and water flows to the surface causing the sand to glisten.

During some quakes a liquefied sand–water mixture is naturally ejected under pressure through cracks in the ground and sand is deposited on the surface. Due to their appearance, these sandy deposits are referred to as 'sand boils'.

When a soil liquefies it completely loses its bearing capacity. Unless a building above it is designed to float like a boat it tilts and overturns (Fig. 7.3). It is usually impractical to design a building to float. The centre of gravity must be lower than the centre of flotation or the centre of gravity of the displaced fluid which, without a heavy 'keel', is unlikely.

If a site is susceptible to liquefaction, and relocation of the proposed building to a less vulnerable site is not an option, what are the alternatives? First, a building can be founded on piles. The objective of piling is to extend the foundations through the liquefiable deposit and found a building on firm soil or a rock stratum unaffected by seismic waves. A limitation of this approach is that the piles will not prevent the ground around and under the building from settling. Unless specifically designed for relative movement, possibly in the order of several hundreds of millimetres, power, water, gas and sewage services are at risk of rupture. The piles themselves are also vulnerable to damage especially due to relative horizontal movements between the ground strata above and below the layer of liquefiable soil. Structural engineers provide these piles with additional ductility.

As an alternative to piling that founds the building on a firm stratum, ground can be strengthened by one of several techniques known as 'ground improvement' (Fig. 7.4).[2] The approaches to ground improvement consist of densification, drainage and reinforcement. Soil can be densified by several different methods. Vibroflotation consists of penetrating the soil with a large vibrating probe, possibly squirting water from its tip. It vibrates and densifies the soil as it sinks into it. In a variation of this method, stones are introduced as the vibrating probe is withdrawn to form a stone column that both reinforces the soil and allows any quake-induced water pressure to dissipate quickly. Stone columns are positioned typically on a one and a half metre grid. A far less sophisticated dynamic compaction method to densify loose soils shallower than approximately 12 m may be an option if there are no

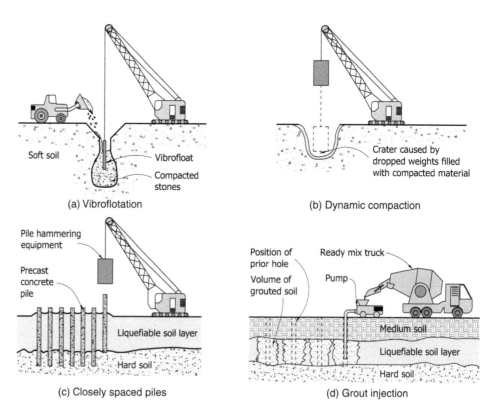

(a) Vibroflotation

Soft soil
Vibrofloat
Compacted stones

(b) Dynamic compaction

Crater caused by dropped weights filled with compacted material

(c) Closely spaced piles

Pile hammering equipment
Precast concrete pile
Liquefiable soil layer
Hard soil

(d) Grout injection

Position of prior hole
Volume of grouted soil
Ready mix truck
Pump
Medium soil
Liquefiable soil layer
Hard soil

▲ 7.4   Common methods of ground improvement to prevent liquefaction.

▲ 7.5 Lateral spreading damage to harbour pavements. 1995
Kobe, Japan earthquake.
(Reproduced with permission from Adam Crewe).

adjacent buildings or facilities. A crane pounds the ground by dropping a weight of up to 30 tonnes from a height of between 10m to 30m. Once the resulting 'craters' are filled with compacted material the building is founded on shallow foundations. At times, contractors dynamically compact soil utilizing an even heavier-handed approach involving blasting. Obviously this technique is limited to sites distant from existing buildings!

The formation of stone columns is one method of improving soils prone to liquefaction and providing a means of drainage. Another soil improvement technique includes driving many closely-spaced piles. This densifies the soil by compressing it and reduces the chance of liquefaction. In permeable soil deposits like sands and gravels, cement or chemical grout can be injected underground at close enough centres in plan to bind the soil particles together and prevent liquefaction.

One of the side affects of liquefaction is lateral spreading – the lateral or horizontal movement of flat or gently sloping reclamations and ground adjoining foreshores and rivers. This has been observed in many earthquakes; it caused extensive damage in the recent 1994 Northridge, California and 1995 Kobe, Japan earthquakes (Fig. 7.5). Depending on various engineering soil parameters, and the intensity of ground accelerations, horizontal displacements can vary from tens of millimetres to several metres.[3] Although lateral spreading may occur in soft soils that do not liquefy, the range of mitigation techniques includes most of those that prevent liquefaction. The construction of sea walls to confine areas likely to flow, and location of buildings well away from free edges likely to undergo lateral spreading, are also recommended.

Subsidence

Earthquake shaking has the same effect on dry and wet loose soils as a vibrator has on fresh concrete – densification. Soil subsides as it densifies, and often unevenly. As well as the risk of a building tilting or suffering severe distortions from this 'natural' process of densification, connections of underground services to a building at foundation level are vulnerable to damage. Damage resulting from this type of differential subsidence can be avoided by using one of the liquefaction mitigation methods.

▲ 7.6 A landslide caused by the 2005 Kashmir earthquake. (Reproduced with permission: US DoD Photo).

Landslides

Dynamic inertia forces within a mass of sloping soil or rock cause slope instability when they exceed the soil strength. During the 2005 Kashmir earthquake large areas of steep terrain suffered from landslides that swept away entire villages (Fig. 7.6). Landslides devastate buildings located upon or below them and can undermine construction built above.

Many different slope stabilization methods are available, including drilling into a slope to drain potential slip surfaces, and installing rock or ground anchors, perhaps prior to shotcreting the soil surface.[4,5]

Surface fault rupture

Any existing structure unfortunate enough to be located above a ruptured fault line is likely to suffer damage. Its severity depends on the magnitude of horizontal and vertical movement along the fault as well as the construction of the building. A low-rise light-weight building is far more likely to survive than one that is heavy and brittle. The likelihood of this type of damage is avoidable by identifying the location of active faults and building away from them. Avoidance can also be enforced by planning regulations, as discussed in Chapter 15. Their aim is to prohibit or limit construction over an active fault, preventing possible catastrophic damage when movement occurs. However, any building, no matter how well-tied together at foundation level and how light-weight, cannot expect to be habitable after significant vertical movement across a fault.

Foundation failure

Inertia-induced overturning moments generate increased compression and possibly tension forces in the end columns of a moment or braced frame and in the chords of a shear wall (Fig. 7.1). If the foundations have been under-designed for the magnitude of those forces, foundation

▲ 7.7 Foundation settlement of up to 500mm contributed to the building damage. 1999, Izmit, Turkey earthquake. (Sezen, H. Courtesy of the National Information Service for Earthquake Engineering, EERC, University of California, Berkeley).

failure occurs. Footings or piles fail when they are pushed excessively into the soil and displace the soil around them (Fig. 7.7). Piles can be pulled from the soil as a building topples over.

It is possible for some buildings to survive an earthquake by rocking. Once a building starts to rock, its natural period of vibration increases and this reduces its acceleration response. Sometimes, especially in shear wall buildings requiring seismic retrofitting, the acceptance of, and design for foundation rocking justifies less invasive strengthening. The Hermès building in Tokyo has been designed to rock, as explained in Chapter 2.

Engineers prevent foundation failure by assessing levels of design forces and sizing the foundations accordingly. The safest method of foundation design uses the Capacity Design approach. Foundation failure is avoided if foundations are designed stronger than the structural fuses in superstructure members.

FOUNDATION TYPES

Figure 7.8 illustrates the most common foundation types in buildings designed for seismic resistance. Table 7.1 elaborates upon each.

Since each foundation system reacts differently to seismic forces, engineers recommend that for a given building foundation systems should not be mixed. If highly variable ground conditions warrant different foundation types over a building plan the impact of their relative stiffness on horizontal force paths needs special attention.

FOUNDATION INVESTIGATIONS

Foundation investigations are generally undertaken in several stages. After resolving the building size or massing, a preliminary investigation takes a broad and general approach similar to that undertaken during site selection. Its focus extends beyond the site to identify potential hazards due to landslides above or below, liquefaction and lateral spreading. Also at this stage the site development history is reviewed, the seismicity of the site assessed, and geological and hazard maps consulted to understand the general geology and check for any active faults crossing through or near the site. Some codes require flexible buildings close to a designated 'active fault' to be designed stronger in anticipation of intense localized earthquake shaking known as 'near-fault fling'. The level of site seismicity influences the requirements for specific seismic soil investigations. In a region of low seismicity, earthquake induced liquefaction or landslides may not be a consideration.

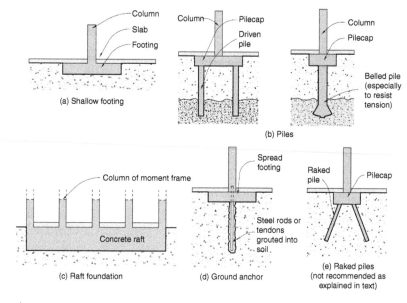

▲ **7.8**   Common foundation types for seismic resistance.

▼ **7.1**   Foundation types and their suitability to resist seismic forces

Foundation type	Comments
Shallow footing	Spread or strip footings are the foundations of choice for low-rise buildings due to their ease of construction but they are unable to resist tension forces greater than their self-weight. Individual footings should be interconnected with tie-beams or a structural slab to prevent any relative horizontal movement occurring during earthquake shaking.
Piles	Most piles are designed to resist compression and tension forces. The amount of tension-resisting friction between their shafts and the ground depends upon the method of installation. Belled piles or piles socketed into bedrock can also provide tension capacity. Piles penetrate soft layers to provide adequate bearing at depth. Tie-beams between piles as for shallow footings are recommended.
Raft	Suitable for medium to high-rise construction. A raft, whose depth can exceed 2 m, integrates gravity-only and seismic resisting vertical members. It mobilizes the entire weight of the building to resist inertia-induced overturning moments. It spreads concentrated loads onto a larger area and makes the structure tolerant of minor ground subsidence.
Ground anchors	Sometimes used in conjunction with shallow footings, ground anchors are an efficient method for resisting tension forces resulting from overturning. They are also suited to anchoring foundations to steep slopes. Particular care against corrosion by using double-corrosion protection is required.
Raked piles	While this system is very stiff against horizontal forces due to its triangulation, it has performed poorly in past earthquakes. Its rigidity and inherent lack of ductility mean raked piles should usually be excluded from seismic-resistant foundations.

Having completed the first and preliminary phase of investigations, reported back, sought and received client approval, more thorough investigations begin. Commensurate with the scale and importance of the project and the geological complexity of the site a range of tests are undertaken. For a small building simple manual penetrometer tests and boreholes may suffice to determine the soil conditions and identify any matters for concern. For a larger project, trenches or machine-drilled *in situ* tests enable engineers to quantify the engineering properties of the soil (Fig. 7.9). Recovered soil samples are subject to a battery of tests in a civil engineering laboratory. For large buildings additional foundation investigations are undertaken at the design stage, and even during construction. Boreholes might be necessary to prove founding conditions are adequate under major piles.

▲ **7.9** A drilling crew undertaking foundation investigations. Wellington, New Zealand.

RETAINING STRUCTURES

Retaining structures located in seismic regions are subject to static soil pressures that are increased by dynamic effects. Where a structure also retains water, or any other fluid for that matter, the structural design process includes hydrodynamic pressures.

Retaining walls

Various types of retaining walls resist horizontal soil pressure (Fig. 7.10). Gravity walls rely on their self-weight to resist sliding and overturning actions from the soil. Reinforced-soil walls use the weight of backfill to mobilize friction forces along buried reinforcement either consisting of long corrosion-protected steel strips or plastic geogrids. They connect to precast facing panels or modular blocks that the compacted backfill bears against. Tieback walls withstand horizontal pressures by ground anchors grouted into the soil or rock behind any potential slip surfaces. Alternatively, natural ground can be reinforced by soil nailing. Soil nails are similar in principle to ground anchors but they comprise more closely-spaced single grouted steel rods that are not prestressed.

The resistance mechanism of cantilever walls is quite different. The horizontal thrust of the soil induces bending moments and shear forces in the relatively thin walls. They cantilever above their foundations that transfer the bending and shear actions to the soil beneath them. If the wall consists of a series of vertical ribs or posts infilled with secondary horizontal-spanning structure like

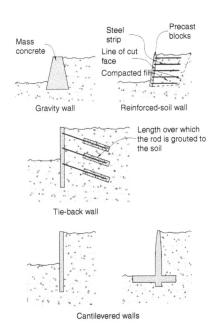

▲ **7.10** Common types of retaining structures.

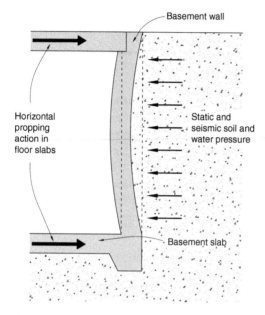

Basement wall

Horizontal propping action in floor slabs

Static and seismic soil and water pressure

Basement slab

▲ **7.11** A reinforced concrete basement wall propped against horizontal soil pressures by slabs, top and bottom.

wood lagging, posts extend into the soil in the same way that a fence post is embedded to resist horizontal force. Alternatively, a wide base consisting of a 'toe' and 'heel' prevents the wall from overturning or overstressing the soil it rests upon. A shear key projecting below the wall base may be required to prevent sliding.

Basements

Whereas cantilever walls by definition cantilever about their bases, basement walls are usually propped against horizontal soil pressure by both basement slabs and suspended floors at the top of the walls (Fig. 7.11) Because of the propping action at ground floor level, the thickness of a basement wall can be considerably less than that of a cantilever wall of the same height, and there is no advantage in tapering its elevational section. As well as resisting out-of-plane static and dynamic soil and water pressures, basement walls often conveniently function as deep and strong foundation beams to stabilize perimeter shear walls or moment frames that rise up from them.

Underground structures

By virtue of being buried, underground structures are subject to substantial vertical and horizontal static pressure. Because of their confinement by surrounding soil, which reduces their dynamic response as compared to similar structures constructed above ground, some engineers thought underground structures were safe during earthquakes. That is, until the 1995 Kobe earthquake. Severe damage occurred to underground subways and other buried facilities when they were deformed by the modest horizontal shear movements within the encompassing soft soil.[6] Now that the seismic damage mechanism is better understood and future brittle fractures can be avoided by ductile design, satisfactory seismic performance of new underground structures can be expected.

Underground structures are vulnerable to flotation in the event of soil liquefaction. If built below the water table their tendency to float to the surface increases. Where liquefaction may occur, underground structures need strong anchorage to dense soil or rock unaffected by seismic shaking, or the ground around them improved by densification and drainage.

REFERENCES AND NOTES

1 Eisner, R. (2006). Site evaluation and selection. In *Designing for Earthquakes: a Manual for Architects FEMA 454*. Federal Emergency Management Agency, pp. 3.1–3.31.

2 Refer to a geotechnical text, such as Kramer, S.L. (1996). *Geotechnical Earthquake Engineering*. Prentice Hall, for more details.

3 Palmer, S. (2006). Assessment of the potential for earthquake induced lateral spreading. *Proceedings of the 2006 Conference of the New Zealand Society of Earthquake Engineering*, Napier, Paper No. 32, pp. 8.

4 Shotcrete is concrete pumped through a hose and nozzle with such a velocity that it can be sprayed onto a surface to form a layer of concrete of the desired thickness.

5 Abramson, L.W. et al. (2002). *Slope Stability and Stabilization Methods* (2nd edn.). John Wiley & Sons.

6 Kawashima, K. (2000). Seismic design of underground structures in soft ground: a review. In *Geotechnical Aspects of Underground Construction in Soft Ground*, Kusakabe, Fujita and Miyazaki, Balkema, pp. 3–20.

8 HORIZONTAL CONFIGURATION

INTRODUCTION

Chapters 4 and 5 introduced readers to the range of horizontal and vertical structural systems found in earthquake-resistant buildings. Each building requires a horizontal system that resists and then distributes inertia forces into the vertical structure (for instance shear walls) provided in a given direction. To account for directionally random shaking, vertical structure is provided in each of two plan orthogonal axes of a building and individual vertical elements are off-set from each other to resist torsion. These are the basic essentials of seismic resistance applicable to buildings.

'Configuration' describes the layout of structure both in plan and elevation. The term encompasses a global 3–D appreciation of how structure and building massing integrate to achieve seismic resistance. This chapter and that following describe commonly occurring configuration challenges that architects face and suggest ways to overcome them without excessively compromising architectural design objectives.

As compared to the seismic performance of individual structural members like beams or columns, building configuration implies a holistic view of a building from a seismic perspective. The effect of building configuration on seismic performance is highlighted after every damaging quake. To a considerable extent the quality of a building's configuration determines, more than many other factors, how well it survives strong shaking. Christopher Arnold, who has written extensively about configuration issues states: 'While configuration alone is not likely to be the sole cause of building failure, it may be a major contributor. Historically, before the use of steel and reinforced concrete construction, good configuration was one of the major determinants of good seismic performance'.[1]

Architects are primarily responsible for building configuration. They determine the overall form or massing of a building and, with or without input from structural engineers, determine the structural layout to suit building function and space planning requirements as well as to express their architectural concepts. Given the importance of architects having a sound appreciation of good and poor seismic configuration, this chapter focuses upon horizontal configuration; namely the floor plan geometry of a building as well as its structural layout in plan. Chapter 9 covers vertical configuration.

Configuration issues, perhaps more than any other, highlight the contrasting aspirations of the architectural and structural engineering professions. Architects are inclined, not surprisingly, to shun ubiquitous rectilinear geometry in favour of more stimulating plan forms. They, and invariably their clients and critics, are attracted towards L, C, or other plan shapes. Geometrical plan complexity can also include large floor penetrations. These might be desirable for introducing daylight or accentuating spatial variety or hierarchy. Such moves away from regular configuration are also motivated by many other factors including: the architectural design concept, the response to the site, the building program, the desirability of introducing natural light and ventilation, and exploiting potential views.

Structural engineers, on the other hand, take a completely different approach towards plan regularity. They adopt the KISS Principle – Keep It Simple and Symmetrical – preferring floor plans as well as structural layout to be as regular and symmetrical as possible. Code requirements, mostly authored and read by structural engineers, reinforce these values. As the European seismic code reminds us: 'To the extent possible, structures should have simple and regular forms both in plan and elevation. If necessary, this may be realized by subdividing the structure into dynamically independent units'.[2]

In their quest for regularity, engineers approach configuration irregularities with the aim of minimizing or eliminating them. One point of potential conflict between the professions might be when an engineer refuses a commission where an architect is unwilling to agree to a more regular horizontal layout. No doubt the architect then shops around for another engineer willing to take a more creative or positive approach towards irregularity. Sadly, the architect may find an engineer less aware of the dangers of poor configuration during a quake.

With greater or lesser degrees of preciseness, codes provide definitions of irregularity. For the purpose of guiding structural engineers on how to approach the design of horizontally irregular structures, one code lists and defines five types of horizontal irregularities in order to classify a building either regular or irregular:

- Torsional and extreme torsional
- Re-entrant corner
- Diaphragm discontinuity
- Out-of-plan offsets, and
- Non-parallel systems.[3]

While this categorization may not impact directly upon an architect, its implications for his or her structural collaborator are far more serious. Irregularity means a far more time-consuming period of design and consequent increase in design costs. Whereas regular structures may be designed by simple and straightforward methods, irregular structures necessitate far more sophisticated approaches. Usually the structural engineer constructs a complete 3–D computer model of an irregular structure before subjecting it to code-specified seismic forces. And even such a complex analytical process, whose accuracy is limited by uncertainties inherent in modelling assumptions, cannot guarantee perfect seismic performance. Based on observations of quake-damaged buildings, experienced engineers acknowledge that the performance of buildings with irregular horizontal configuration is unlikely to be as good as that of more regular structures.

Irregularity leads to other structural disadvantages. Codes may require structural connections and members to be stronger than normal and therefore more expensive. Codes may also penalize irregularity by requiring larger design-level forces. One code requires torsionally irregular structures to be designed 50 per cent stronger.[4] But the ultimate penalty for irregularity is the withholding of permission to build. This is the case where at least one code prohibits irregular structures being built in regions of high seismicity.[5]

In the context of Venezuela, researchers quantified the life-time costs and benefits of horizontal irregularity of two identically sized hotels. One was rectilinear and the other L-shaped to better suit the site and its surrounds.[6] The benefits of irregularity included increased natural light, reduced mechanical ventilation in some areas and, most significantly, increased income from rooms with better ocean views. Benefits far outweighed the costs of irregularity such as increased structural costs to achieve adequate seismic performance and the increased construction cost associated with more complex geometry. The researchers mitigated the anticipated more severe seismic damage in the irregular structure by increased structural analysis and design effort that resulted in greater structural member sizes. Had they taken the path of subdividing the structure into seismically independent units, and included the life-time reduction in seismic damage costs associated with regular structures and the additional costs of separation as discussed later, the financial outcome might have even improved.

While acknowledging architects' attraction to innovative forms that, among other reasons, respond to their sites and express architectural ideas, the following sections of this chapter examine specific potential horizontal configuration problems. The challenges presented and ways to minimize them are noted. Although one can imagine extreme configurations that would be impossible to design and build safely, less poorly configured buildings can usually be resolved satisfactorily by adopting the approaches suggested. As will become evident, the most important strategy for achieving satisfactory horizontal configuration is by simplifying complex building plan shapes by separating regular shapes from each other with seismic separation gaps. The chapter concludes by elaborating upon the architectural and structural consequences of horizontal separation.

TORSION

Torsion was introduced in Chapter 2. In summary, if the Centre of Mass (CoM) of a building is not coincident with the Centre of Resistance (CoR) a torsional moment acts in the horizontal plane causing floor diaphragms to twist about the CoR (see Fig. 2.16). The rotation affects columns located furthermost from the CoR most severely. They are subject to large horizontal deflections, sometimes damaging them so seriously they collapse under the influence of their vertical gravity forces. Numerous torsion failures were observed during the 1994 Northridge and 1995 Kobe earthquakes (Fig. 8.1). Based upon post-earthquake observations of building failures, torsion is recognized as one of the most common and serious horizontal configuration problems.

Architects and engineers prevent building damage arising from torsion by using several approaches. Firstly, they minimize the distance in plan between the CoM and CoR. Remember that even with a perfectly symmetrical structural configuration some degree of torsion still occurs due to torsional motions within the ground shaking. Codes specify a minimum design eccentricity to account for this and unavoidable out-of-balance or asymmetrical distribution of gravity forces in a building with respect to the CoR. *Every*

▲ **8.1** Collapse of a concrete frame building at ground floor level due to torsion, 1995 Kobe, Japan earthquake. (Reproduced with permission from EERI. David R. Bonneville, photographer).

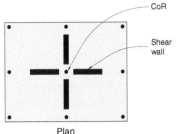

Plan
(a) No torsional resistance

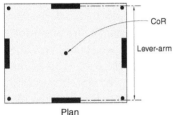

Plan
(b) Excellent torsional resistance

▲ 8.2 Two structural configurations, each with four shear walls, with contrasting abilities to resist torsion.

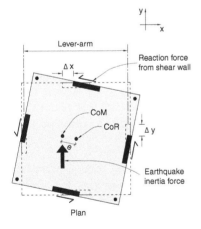

Plan

▲ 8.3 A building plan illustrating how vertical structure resists torsion. Most gravity-only structure and the movement of the building in the y direction is not shown.

building, no matter how symmetrically configured in plan, requires torsion resistance.

Secondly, designers provide a minimum of two lines of vertical structure parallel to each of the main orthogonal axes of a building yet horizontally offset from each other. The horizontal off-set or lever-arm between each line of structure should be as large as possible to maximize both the latent torsion-resisting strength and stiffness. When the building in Fig. 8.2(a) twists in plan, its shear walls offer no significant resistance because they warp, flexing about their weak axes. In contrast, when the plan in Fig. 8.2(b) twists about the CoR which is centrally located, each of the four walls reacts along its line of strength against the horizontal deflection imposed upon it by the rotation of the floor diaphragm. Long lever-arms between pairs of walls provide the best possible resistance against torsion.

How exactly does vertical structure resist torsion? Consider the building in Fig. 8.3. It is very well configured structurally to resist torsion – two perimeter shear walls in each direction. Assuming a torsional eccentricity e between the resultant line of action of inertia forces acting in the y direction and the CoR, the building twists clockwise. Its diaphragm rotates as a rigid unit. A diaphragm is usually very stiff and strong in its plane, especially if constructed from reinforced concrete. When twisting occurs about the CoR, which is the point through which the resistance from all the shear walls acts, the shear walls acting in the y direction deflect in opposite directions a small amount Δy. These movements are additive to the shear wall deflections due to the y direction forces that are not shown. Each shear wall also twists a little. This source of torsional resistance is neglected because the twisting strength of an individual wall is so low. As each wall is pushed, it resists the imposed deflection in the direction of its strength (the y direction) and applies a reaction force. The value of these reaction forces multiplied by the lever arm between them represents a moment couple that partially resists the torsional moment causing diaphragm rotation.[7]

Also due to the diaphragm rotation, the x direction shear walls deflect horizontally Δx in opposite directions. Like the y direction shear walls, they react against the movement that deflects them. They apply equal and opposite reaction forces upon the diaphragm creating another moment couple. Even though no x direction seismic forces act on the building, because these two shear walls orientated parallel to the x axis are strongly connected to the diaphragm, they nonetheless participate in resisting torsion. The two torsion-resisting couples formed by the

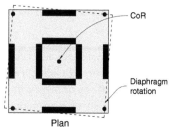

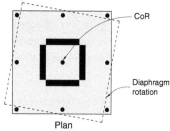

(a) Four inner walls slightly
increase torsional resistance

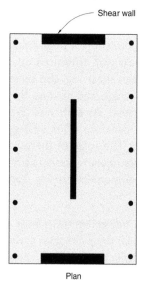

(b) Diaphragm rotation increases
where inner walls alone resist torsion

▲ **8.4** Structure located close to the CoR
is less effective at resisting torsion.

Plan

▲ **8.5** An example of a torsionally
unbalanced system.

pairs of parallel shear walls combine to resist the torsional moment and provide torsional equilibrium. Any structural damage is unlikely since only minimal diaphragm rotation occurs.

The four extra shear walls added in plan (Fig. 8.4(a)) enhance torsional resistance slightly. Even if the new walls are identical to the perimeter walls because they are closer to the CoR they are subject to 50 per cent smaller displacements when the diaphragm twists and the lever-arms between them are less. With a lesser resisting force (proportional to horizontal displacement) and half the lever arm their torsion-resisting contribution is only 25 per cent of that provided by the perimeter walls. If the perimeter walls are removed, and horizontal forces and torsion are now resisted by the inner walls alone, the two torsion-resisting couples must offer the same resistance as before since the value of the torsion moment is unchanged. We can neglect any torsional resistance from the slender perimeter columns. Since the lever-arms between the inner walls are half of the original lever-arms wall reaction forces double. This means that these walls will need to be considerably stronger and that the diaphragm will twist further. The structural configuration in Fig. 8.4(b) is therefore twice as torsionally flexible as that in Fig. 8.2(b); but it might still be structurally adequate especially if the perimeter gravity-only columns can sustain the ensuing horizontal movements without damage.

Although the previous figures illustrate shear walls resisting seismic forces, moment and braced frames can also provide adequate torsion resistance. Replace the shear walls with one- or multi-bay moment frames and the principles outlined above still apply. The building will be less torsionally stiff due to the lesser stiffness of the frames but still perform adequately, especially if the frames are located on the perimeter of the building.

In the examples considered so far, a recommended torsion-resistant structure comprises a minimum of four vertical elements, like shear walls or moment frames, with two in each direction. However, in some situations the number of elements can be reduced to three (Fig. 8.5). Any y direction forces are resisted by one shear wall, albeit long and strong especially given an absence of redundancy, and x direction forces resisted by two walls. When torsion induces diaphragm rotation, the two x direction walls, in this case with a long lever-arm between them, form a moment couple. They provide torsional stability or equilibrium irrespective of the direction of loading – but only so long as they remain elastic. Most shear walls and frames are designed for relatively low seismic forces if they incorporate ductile detailing. So when one x direction wall yields as a result of inertia forces in the x direction as well as torsion it temporarily loses its stiffness and the COR migrates towards the stiffer end, increasing torsional eccentricity. The system becomes torsionally unstable.

This configuration consisting of three vertical structural elements is described by structural engineers as a *torsionally unbalanced system*. It is not recommended unless the x direction walls or frames are much stronger than minimum requirements. They must be capable of resisting horizontal forces with little or no ductility demand and therefore possibly possess far more strength than normal. Researchers are currently responding to the undesirable situation where code torsion provisions are based upon elastic structural performance and have yet to account for the effects of anticipated inelastic or ductile behaviour.[8] Until research findings update guidelines, architects should avoid torsionally unbalanced systems unless satisfying the criterion above.

The beginning of this section recommended that designers minimize torsional eccentricity. But to what extent? When is torsional eccentricity too great? A structural engineer can provide an answer for a particular building, but only after undertaking a complex 3–D computer analysis to calculate dynamic stresses and horizontal displacements. As a rule-of-thumb, keep the eccentricity in each orthogonal direction to less than 25 per cent of the building dimension measured normal to the direction of force under consideration.

The worst-case torsion scenario, as shown in Fig. 8.6, is common for buildings on urban corner sites. The two sides away from the street are bounded by fire-resistant walls which, even if not specifically designed as shear walls, act as such. The two street frontages are relatively open. The CoR is therefore located at the back corner of the building. Assuming the building weight is evenly distributed in plan the eccentricities in both directions equal 50 per cent of the building plan dimensions.

If you design a corner building use one of three strategies to reduce torsional eccentricities. First, avoid designing strong rear walls. Substitute them with fire-resistant infilled moment frames. Infills of either lightweight construction or reinforced masonry that are separated on three sides from the frames, as discussed in Chapter 10, are suitable. Both those frames – which also carry gravity forces – require identical frames along the street frontages to balance them torsionally (Fig. 8.7).

Secondly, in an alternative, but less popular strategy, the two strong rear walls remain, but are separated in-plan from the rest of the building

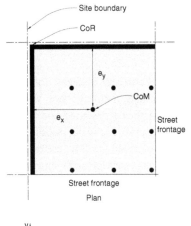

▲ **8.6** Poor configuration of a typical corner building with shear walls as boundary walls.

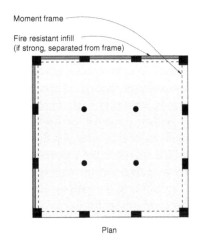

▲ **8.7** An improved horizontal configuration for a corner building. Horizontal forces are resisted by symmetrical moment frames.

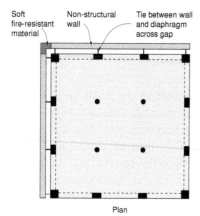

▲ **8.8**   An alternative approach to achieving satisfactory configuration of a corner building by separating the strong walls from the rest of the structure and providing new moment frames to resist all seismic forces.

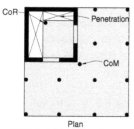

Plan
(a) Undesirable configuration caused by an eccentric structural core

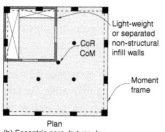

Plan
(b) Eccentric core, but regular horizontal configuration

▲ **8.9**   Avoiding torsional eccentricity from an eccentric structural core by making the core non-structural.

(Fig. 8.8). Horizontal separation joints between the walls and floor diaphragms prevent the walls playing any role in seismic resistance. New moment frames just inside the walls resist inertia forces in both directions. The final step is to detail the wall-to-diaphragm connections at each level. Ties must resist inertia forces acting normal to the walls arising from their self-weight. The connections need to be strong along one axis, but able to slide or move freely along the other. The associated detailing complexities explain why this is a less preferred option.

The third strategy for reducing torsional eccentricity involves 'softening' the rear walls by designing and constructing them as many short walls. For example, a single 20 m long wall might be designed and built as five 4 m long walls. The vertical joints between the walls, necessary for ensuring structural separation, require treatment for fire resistance. This approach needs to reduce torsional eccentricity sufficiently to make it attractive. Even softened walls can be far stronger and stiffer than moment frames.

The strategy of substituting potentially strong walls with infilled frames or constructing the walls from light-weight and relatively weak but fire-resistant materials, thereby rendering them non-structural, is often a viable solution to reduce torsion. Its usage is not confined solely to the perimeters of buildings. It can overcome eccentricities associated with, for example, eccentrically placed cores or shafts that might normally function as horizontal force-resisting elements (Fig. 8.9).

RE-ENTRANT CORNERS

Buildings that have suffered seismic damage due to re-entrant corners occasionally feature in earthquake reconnaissance reports. Although re-entrant geometries can take many shapes, what they share in common from a seismic design perspective, is their potential for damage resulting from the different dynamic properties of each wing (Fig. 8.10). For example, when the building in Fig. 8.11 is shaken in the y direction, the left-hand area of the building, and the wing to the right, react quite differently. The left-hand area deflects horizontally a relatively small amount due to its greater depth and inherently greater horizontal stiffness. The more flexible wing moves further and at a different period of vibration. It swings about the stiffer area, possibly damaging floor diaphragms at the junction of the two wings. As a result of the large horizontal deflections, the right-hand end columns of the right-hand wing might also sustain damage. Effectively, the right-hand wing is subject to

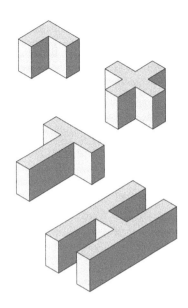

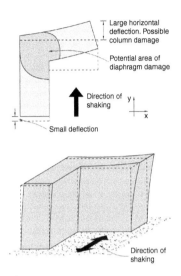

Large horizontal
deflection. Possible
column damage

Potential area of
diaphragm damage

Direction of
shaking

Small deflection

Direction of
shaking

▲ **8.11** The dynamic response of a
re-entrant configuration and potential
floor diaphragm damage area.

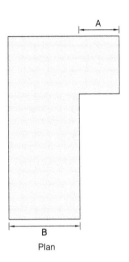

Plan

▲ **8.12** A typical
definition of an irregular
re-entrant configuration is
where A > 0.15B.

▲ **8.10** Typical re-entrant corner forms.

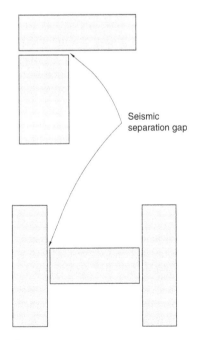

Seismic
separation gap

▲ **8.13** Irregular plan configurations
improved by seismic separation gaps.

torsional rotation about the stiffer and stronger left-hand area. Shaking in the *x* direction highlights the same configuration problem.

The attitude of most codes towards re-entrant corners is to require structural engineers to undertake a 3–D dynamic analysis where the length of a projecting area of building causing a re-entrant corner exceeds approximately 15 per cent of the building plan dimension (Fig. 8.12). An engineer will design the re-entrant structure to avoid either diaphragm tearing or excessive horizontal deflections. This can be achieved by fine-tuning the relative stiffness of the wings. However, if they are long or their diaphragms weakened by penetrations for vertical circulation or other reasons in the critical region where they join, that approach may not be structurally sound. The building might best be separated into two independent structures.

Separation is a common solution for re-entrant corner buildings (Fig. 8.13). How it is achieved is explained later in this chapter. Although a building might be perceived as a single mass if its blocks are seismically separated, it actually consists of two or more structurally independent units, each able to resist its own inertia forces including torsion. Where possible, separation gaps are provided adjacent to, or through, areas where floor diaphragms are penetrated or discontinuous.

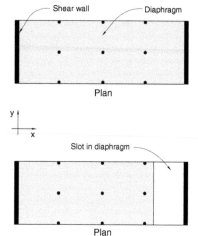

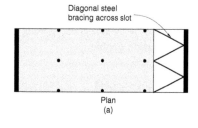

▲ 8.14 A slot in the diaphragm destroys its ability to span between shear walls for y direction forces. (X direction structure not shown.)

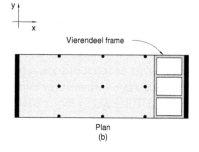

▲ 8.15 Plans of two diaphragms where structural integrity across a slot is restored by steel bracing and frame-action.

DIAPHRAGM DISCONTINUITIES

In the ideal world of the structural engineer, diaphragms in buildings are not penetrated by anything larger than say a 300 mm diameter pipe. Diaphragms are also planar and level over the whole floor plan. However, the real world of architecture is quite different, because in most buildings quite large penetrations are required for vertical circulation such as stairways and elevators. Building services, including air ducts and pipes also need to pass through floor slabs and in the process introduce potential weaknesses into diaphragms.

Chapter 4 outlines the roles and requirements of diaphragms. It likens diaphragms to horizontal beams resisting and transferring horizontal inertia forces to their supports which, in this case, consist of vertical structural systems such as shear walls or moment frames. It explains how penetrations are acceptable structurally, provided they respect the shear force and bending moment diagrams of a diaphragm. Recall that the web of a diaphragm resists shear forces, while perimeter diaphragm chords acting in tension or compression, resist bending moments.

The size of a penetration can be large enough to ruin the structural integrity of a diaphragm altogether. Consider the case of a simple rectangular diaphragm spanning between two shear walls that act in the y direction (Fig. 8.14). What are the structural options if a full-width slot is required? The slot destroys the ability of the diaphragm to span to the right-hand wall. If the purpose of the slot is to introduce light or services through the diaphragm one option is to bridge the slot by introducing a section of steel bracing (Fig. 8.15(a)). If designed and connected strongly enough it restores the original spanning capability of the diaphragm. Alternatively, if the geometry of diagonal members isn't acceptable aesthetically a horizontal vierendeel frame, with its far larger member sizes, can be inserted to restore structural function (Fig. 8.15(b)). In both solutions, light and services can pass between structural members.

If the intention of the penetration in Fig. 8.14 is to provide a staircase, then both previous options are unacceptable. It is now impossible for the diaphragm to transfer forces to the right-hand shear wall. The only option is to no longer consider that wall as a shear wall but to provide a new shear wall to the left of the penetration. Now a shortened diaphragm spans satisfactorily between shear walls. The force path has been restored. All that remains to complete the design is to stabilize the right-hand wall for x direction forces by tying it back to the newly

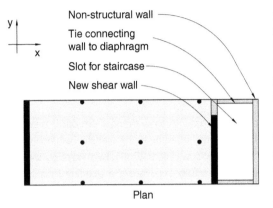

Non-structural wall

Tie connecting
wall to diaphragm

Slot for staircase

New shear wall

Plan

▲ **8.16** A new shear wall enables the right-hand wall to become non-structural.

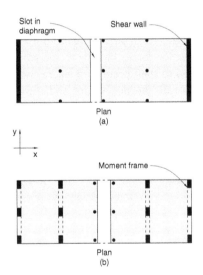

Slot in diaphragm

Shear wall

Plan
(a)

Moment frame

Plan
(b)

▲ **8.17** A diaphragm slotted near the middle leads to the formation of two separated structures (a). To avoid serious torsional eccentricities, the shear walls are substituted by moment frames (b). The torsional configuration of each structure can be improved if the inner two frames are moved closer to the gap. (X direction structure not shown.)

down-sized diaphragm (Fig. 8.16). The two new ties may also have to act as horizontal cantilever beams or members of a vierendeel frame. This will transfer seismic forces from the now non-structural wall to the diaphragm if there is insufficient bracing in the wall to deal with its own inertia forces.

Figure 8.17 considers a more difficult scenario. Now a penetration is required near the middle of a diaphragm, also spanning between two walls. If the insertion of any horizontal structure like the diagonal bracing of Fig. 8.15(a) is impossible due to architectural requirements the only option is to physically separate the two portions of the building. Although perhaps perceived as one building with penetrated diaphragms, each section now becomes an independent structure. The end shear walls need to be replaced by moment frames to minimize torsion (Fig. 8.17(b)). All non-structural connections bridging the gap, such as walls and roof, are detailed to accommodate the relative seismic movements between the two structures.

Another equally serious diaphragm discontinuity occurs where a potential floor diaphragm consists of more than one level. If a relatively small area is raised or lowered it can be treated, as far as seismic behaviour is concerned, as if it were a penetration. But consider the situation where a step is introduced across a diaphragm near the middle

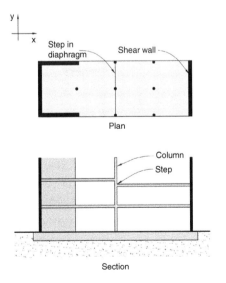

Step in diaphragm

Shear wall

Plan

Column

Step

Section

▲ **8.18** A stepped diaphragm.

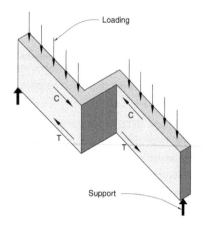

▲ **8.19** A kinked beam showing internal compression and tension forces that can not be achieved. The beam is structurally unsound.

of its span (Fig. 8.18). The diaphragm is now kinked, and just as a beam kinked in plan is unable to transfer force neither can a kinked diaphragm (Fig. 8.19). If you are skeptical, model a simple straight beam from cardboard. Note how it withstands reasonable force where spanning a short distance. Now introduce a kink. Observe how you have destroyed the integrity of the beam.

The other problem caused by the step is to prevent x direction inertia forces from the right-hand end of the building being transferred into the two shear walls acting in that direction (Fig. 8.20(a)). Two ways to overcome these problems are; firstly, to fully separate the building into two structures as discussed previously; or secondly, to introduce a shear wall or frame along the line of the step (Fig. 8.20(b)) and provide x direction shear walls at each end of the building. Now there are two diaphragms. Both span independently between their original perimeter lines now braced by moment frames and a new frame along the line of the step. Frames have replaced the walls to allow for circulation between both halves of the floor plan. If the step is higher than several hundred millimetres, one diaphragm will apply y direction forces directly to the columns of the centre frame. This could lead to their premature failure and so the best approach would be to separate the diaphragms and their supporting members into two independent structures.

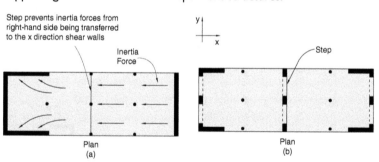

▲ **8.20** The structural difficulty posed by the diaphragm step (a) is solved by increasing the number of shear walls effective in the x direction to four and connecting two to each diaphragm section (b). Moment frames replace y direction shear walls to avoid a mixed system once a moment frame is introduced along the step. Had a shear wall been introduced along the step, the original shear walls in the y direction could have remained.

NON-PARALLEL SYSTEMS

Figure 8.21 illustrates two non-parallel systems. In each case the directions of strength of the vertical structures are angled with respect to any sets of orthogonal axes. The ability of each configuration to resist horizontal forces and torsion is understood by considering the length of each vertical system as a strength vector. A vector can be resolved

▲ **8.21** Two examples of non-parallel systems. Gravity-only structure not shown.

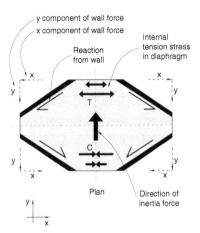

▲ **8.22** A non-parallel system showing the orthogonal force components of each wall and secondary diaphragm stresses for a y direction force.

into components parallel to, and normal to, a set of axes (Fig. 8.22). But what is less apparent is that when these systems resist horizontal force their skewed orientation leads to unexpected secondary forces that are required to maintain equilibrium. In this symmetrically configured building, as the shear walls resist y direction forces, the diaphragms must provide tension and compression forces to keep the system stable. When the configuration of non-parallel systems is asymmetrical the distribution of these internal forces becomes far more complex. For this reason codes insist that structural engineers model non-parallel systems in 3–D in order to capture these effects and design for them.

POUNDING AND SEPARATION

Buildings pound each other during an earthquake where seismic separation gaps between them are insufficiently wide (Fig. 8.23). Due to the damage caused by pounding in past earthquakes, modern codes require designers to provide separation gaps of adequate width. As mentioned in Chapter 1, the need for these gaps means locating buildings back from all site boundaries except street frontages. A building must not drift across a site boundary and damage its neighbour (Fig. 8.24).

The widths of seismic separation gaps depend upon the flexibility of a building and its height. Consider a very flexible building. Assume it is designed to a typical code maximum-allowed seismic drift equal to 0.02 × height. If the building is 10 storeys high with an inter-storey height of 3.5 m the required separation gap at roof level is 700 mm. In theory the gap width can taper from zero at ground level to 700 mm at the roof but this leads to obvious detailing and construction problems. Some codes allow the gap width to be reduced by 50 per cent if the floor levels of adjacent buildings align. Pounding of floor slabs is less damaging than a floor slab slamming into and damaging the perimeter columns of a neighbouring building.

▲ **8.23** Pounding damage during the 1985 Mexico City earthquake. (Reproduced with permission from David C. Hopkins).

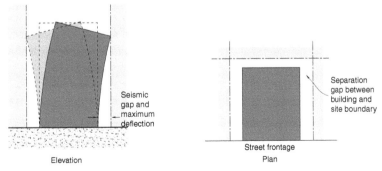

Seismic gap and maximum deflection

Elevation

Separation gap between building and site boundary

Street frontage

Plan

▲ **8.24** Seismic drifts and required separations in elevation and plan to avoid pounding.

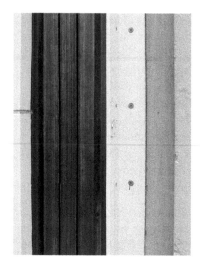

▲ **8.25**   Concertina flashing between two buildings, San Francisco.

As demonstrated above, seismic separation gaps can become quite wide. The only way to minimize them is to design a stiffer structural system. An architect has to decide what is less problematic and more economical – larger vertical structural members or wider separation gaps. Each option yields a different usable floor area. The structural engineer might need to undertake several alternative designs before making a final decision. Mark Saunders calculates that the cost of floor area lost to separation gaps for a 40–storey building in San Francisco approximates 4 per cent of its value.[8] Vertical gaps between separated buildings are usually flashed by a flexible concertina-type detail (Fig. 8.25). At roof level a typical detail allows free horizontal movement in two directions, towards and away from an adjacent building, and to-and-fro parallel to the gap length (Fig. 8.26).

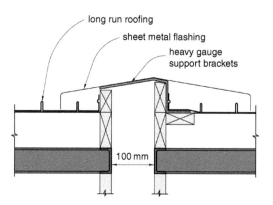

long run roofing

sheet metal flashing

heavy gauge support brackets

100 mm

Section detail at roof separation

▲ **8.26**   Possible detail of a seismic separation gap between two buildings at roof level.

A seismic gap is also required where one building is separated into two independent structures. Detailing must allow for two-dimensional relative out-of-phase movements between blocks. Some codes don't require the seismic gap between the structures to be determined by simply adding the maximum seismic drifts of each block. They allow the two drift values to be combined in such a way that recognizes the low probability of both maximum drifts occurring at precisely the same time. Nevertheless, seismic joints are wide and expensive. The cost can be minimized by having them pass through diaphragm penetrations like stairs and elevators.

Satisfactory architectural treatment of seismic gaps is also required for junctions between floors, walls and ceilings (Figs. 8.27 and 8.28). Architects design and detail flashings and linings to cover the gaps in such a way as to avoid tying the separated structures together. Details must allow movement but it is permissible for them to be damaged in

A = max. movement apart
B = max. movement towards each other

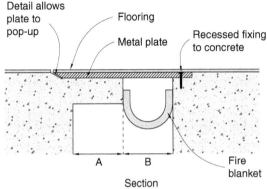

Detail allows plate to pop-up

Flooring

Metal plate

Recessed fixing to concrete

A   B

Fire blanket

Section

▲ **8.27** A section through a generic floor level seismic gap. Dimension 'A' can be reduced if damage caused by the plate sliding off its left-hand seating is acceptable during smaller relative drifts.

▲ **8.28** A seismic joint between floors, walls and ceilings of two separated structures, San Francisco.

moderate shaking as they can usually be replaced easily. They are considered 'sacrificial' in that their damage pre-empts more serious damage elsewhere. Gaps are frequently designed to be fire-resistant and are acoustically treated.

Seismic separation joints also raise various structural issues sometimes with significant architectural implications. For example, how are gravity forces to be supported on each side of a joint? One approach involves providing double beams and columns with the gap running between them (Fig. 8.29(a)). In another method (Fig. 8.29(b)) one floor system cantilevers. It is propped on sliding joints allowing free relative movement by the other structure. Obviously the horizontal overlap of the floor and the supports must allow for the structures to move towards and away from each other and should be dimensioned conservatively. The consequences of a floor slab falling off its supports are severe.

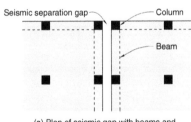

Seismic separation gap   Column

Beam

(a) Plan of seismic gap with beams and columns on either side of the gap

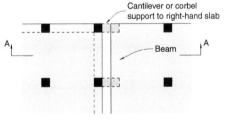

Cantilever or corbel support to right-hand slab

A   A

Beam

(b) Plan of seismic gap with beams and columns on one side

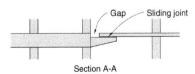

Gap   Sliding joint

Section A-A

▲ **8.29** Two methods of supporting flooring at a seismic separation gap.

▲ **8.30** Collapsed bridge formerly spanning between two buildings. 1995 Kobe, Japan earthquake.
(Reproduced with permission from US National Geographic Data Center. Dr. R. Hutchison, photographer).

BRIDGING BETWEEN BUILDINGS

An extreme example of spanning a seismic gap between independent structures occurs when they are bridged. In urban environments bridges spanning between buildings are a common sight. Designers must ensure that when the buildings move out-of-phase with each other the bridges remain undamaged. The form of bridging may be a literal bridge, providing horizontal circulation from one building to another, or perhaps a glazed roof canopy that creates a sheltered courtyard. The primary design challenge to be overcome is to cope with out-of-phase horizontal movements between the buildings. How can a bridge span between two buildings without connecting them, and how should a bridge be secured to prevent it falling (Fig. 8.30)? If firmly connected at each end a bridge is likely to be torn apart or buckle when the buildings vibrate or drift horizontally relative to each other.

The conceptual starting point of the design is acknowledging that the buildings will move relative to each other during an earthquake. The bridge is therefore anchored or fixed in its longitudinal direction to one building and free to slide on the other. Each end of the bridge must also be restrained at right-angles to its length to prevent it moving sideways under transverse wind and earthquake inertia forces. In Fig. 8.31 the

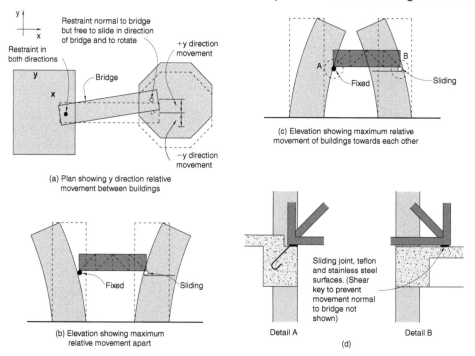

(a) Plan showing y direction relative movement between buildings

(b) Elevation showing maximum relative movement apart

(c) Elevation showing maximum relative movement of buildings towards each other

(d)

▲ **8.31** The relative drifts between two separated buildings (a) to (c) and generic bridge seating details (d).

bridge is connected to the left-hand building and slides on the other in the direction of the bridge length. The seating width at the sliding end should be conservatively assessed to ensure that the bridge will never fall off. Quite sophisticated architectural flashings are required to weather-proof the right-hand bridge-to-building connection. They must be sufficiently flexible to prevent build-up of force when the bridge slides and survive small quakes without repair.

REFERENCES AND NOTES

1 Arnold, C. (1984). Building configuration: the architecture of seismic design. *Bulletin of the New Zealand National Society for Earthquake Engineering*, **17**:2, 83–88.

2 European Committee for Standardization (2004). *Eurocode 8: Design of Structures for Earthquake Resistance – Part 1: General Rules, Seismic Actions and Rules for Buildings (BS EN 1998–1:2004)*.

3 American Institute of Civil Engineers (2006). *Minimum Design Forces for Buildings and Other Structures (ASCE/SEI 7-05)*. American Institute of Civil Engineers. This code satisfies the seismic design requirements of the International Building Code 2006 (IBC 2006), published by International Code Council, Inc.

4 Kitagawa, Y. and Takino, F. (1994). Chapter 24 Japan. In *International Handbook on Earthquake Engineering: Codes, Programs, and Examples*, Paz, M. (ed.). Chapman & Hall.

5 American Institute of Civil Engineers (2006). *Minimum Design Forces for Buildings and Other Structures (ASCE/SEI 7-05)*.

6 López, O.A. and Raven, E. (1999). An overall evaluation of irregular-floor-plan shaped buildings located in seismic areas. *Earthquake Spectra*, **15**:1, 105–120.

7 A moment couple is the moment from two equal and opposite parallel forces. The moment equals the product of the magnitude of one of the forces and the perpendicular distance between them.

8 Paulay, T. (2000). Understanding torsional phenomena in ductile systems. *Bulletin of the New Zealand Society for Earthquake Engineering*, **33**:4, 403–420.

9 Saunders, M.C. (2006). Seismic joints in steel frame building construction. *Practice Periodical on Structural Design and Construction*. American Society of Civil Engineers, **11**:2, 71–75.

9

VERTICAL CONFIGURATION

INTRODUCTION

The vertical configuration of a building encompasses two aspects of architectural form – the building envelop profiles in elevation and the elevation of the vertical structural systems in both orthogonal directions. The best possible seismic performance is achieved where both the 3–D massing and vertical structure of a building are regular. This means an *absence* of the following vertical irregularities repeatedly observed after earthquakes to have initiated severe damage:

- a floor significantly heavier than an adjacent floor
- vertical structure of one storey more flexible and/or weaker than that above it
- short columns
- discontinuous and off-set structural walls, and
- an abrupt change of floor plan dimension up the height of a building.

Most of these irregularities, like the horizontal irregularities discussed in the previous chapter, are described and defined in more or less detail by seismic design codes.[1] Each configuration irregularity modifies the dynamic response of a building and increases structural and non-structural damage. Some minor to moderate irregularities are acceptable to structural engineers and design codes and are dealt with by applying more sophisticated design techniques. The situation of a heavy floor is one such example. Local increases in stresses and inter-storey drifts from one particularly heavy floor, perhaps accommodating a swimming pool, are identified by the structural engineer who then designs the structure using a more exacting procedure than normal. The other irregularities listed above so seriously affect the seismic performance of a building they should be avoided at all cost.

All possible examples of poor configuration cannot be foreseen or addressed in this book. Poor vertical configuration entails force path discontinuities or complexities which affected structural members are generally unable to cope with without severe damage. Wherever the vertical configuration of a proposed structural system varies from the advice of Chapter 5, its seismic adequacy should be queried and then carefully assessed.

The following sections discuss the most frequently observed vertical configuration problems, both from structural and architectural viewpoints. Given that architects may desire a building elevation or an interior structure that *appears* to result in one of these undesirable configurations, methods of achieving architectural design intentions without compromising seismic performance significantly are presented.

SOFT STOREYS

Soft storey configuration describes structure where one storey of a building is more flexible and/or weaker than the one above it from the perspective of seismic forces. Rather than earthquake energy absorbed by ductile yielding of steel reinforcing bars, or structural steel sections in plastic hinge zones, or structural fuses throughout the *whole* structure as shown in Fig. 5.44(b), in a soft storey configuration earthquake energy concentrates on the soft storey (see Fig. 5.44(a)). Serious damage is caused especially to the columns of that soft storey. Once these structural members are damaged the nature of earthquake shaking is *not* to move on and damage other members. Rather, the quake intensifies its energy input and damaging power in that same storey. Often the structure above a collapsed soft storey is virtually undamaged. It has been protected by the sacrificial action of the soft storey. A soft storey building is doomed, since columns in the soft storey usually lack the resilience to absorb seismic damage and still continue to support the weight of the building above.

The sport of boxing provides an apt analogy for soft storey performance. Both boxing and earthquakes are violent. They pummel and injure. A boxer attempts to discover and then exploit an opponent's weakness. Once discovered, that is where the fury focuses and punches land. A 'softening-up' process, whereby injurious blows slow down reactions and lower defences, continues until the knock-out blow. So too with a soft storey. Once a soft storey is found, the quake focuses its harmful attention upon the relatively few vertical elements in that one storey until the building is either staggering or collapses.

▲ **9.1** Soft storey ground floor collapse of a four-storey building. 1995 Kobe earthquake.

Of all vertical configuration problems, the soft storey is the most serious and is by far the most prevalent reason for multi-storey building collapses. So many buildings, located in seismically active regions throughout the world possess relatively open ground floors and are at risk of a soft storey mechanism forming. A report on the 1995 Kobe earthquake observes that ground floor collapse was the most common failure mode in small commercial and mixed-occupancy buildings (Fig. 9.1). Regarding larger commercial and residential buildings, which in most cases appeared regular from the street, the report notes: 'Partial or full collapse of a single story of buildings was the common "collapse" failure mode..... The particular story that sustained partial or full collapse varied from building to building ...'[2] Soft storey collapses are a common occurrence during any strong ground shaking in a built-up region (Fig. 9.2).

Figure 5.44(a) illustrates ground floor soft storey behaviour caused by weak columns and strong beams. Before suffering damage, that storey may not have been any weaker than the storey above but seismic shear forces and bending moments in moment frame columns increase towards the base of a building. They reach their maximum values at the ground floor making it the most vulnerable. In the absence of the Capacity Design approach, columns are usually weaker than beams so columns alone sustain damage. Once 'softening-up' begins during a quake the prognosis is very poor.

Soft storeys are also caused by other configuration irregularities as illustrated in Fig. 9.3. The ubiquitous soft storey caused by a combination of open ground floors and masonry infilled frames (Fig. 9.3(a)) is discussed in detail in Chapter 10. As mentioned above, if the soft storey irregularity is reasonably minor, a seismic code may permit the system to resist horizontal forces. However, the structural engineer must undertake special analyses and provide members within that storey with additional strength and ductile detailing. In more severe soft storey cases even the most advanced structural design cannot prevent poor performance in a design-level earthquake. So the questions arise: 'Is it possible for a building to exhibit the visual characteristics of a soft storey for architectural reasons and still perform satisfactorily in a quake; and if so, how?'

| (a) Stiff and strong upper floors due to masonry infills | (b) The columns in one storey longer than those above | (c) Soft storey caused by discontinuous column |

▲ **9.3** Examples of soft storey configurations.

▲ **9.4** A weak column-strong beam structure develops a soft storey at ground level once columns are damaged.

Fortunately for the aesthetic satisfaction of architects, building users and the public at large who appreciate slender columns and some degree of design variety, the answer is a resounding 'Yes'. One of two strategies is employed: either separation or differentiation. Separation involves isolating from the force path those stiff and strong elements – like infill walls and deep beams – which cause adjacent elements – like columns – to be relatively more flexible and weaker. Differentiation describes a design approach that clearly distinguishes between gravity-only and seismic resisting structure and ensures that selected members primarily resist *either* seismic *or* gravity forces. The following examples illustrate the application of these strategies.

Imagine that you are designing a building whose façade is modulated by slender columns and deep beams (Fig. 9.4). How can

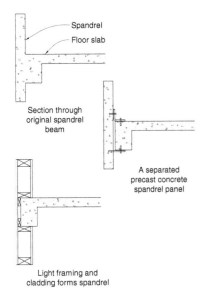

Spandrel

Floor slab

Section through
original spandrel
beam

A separated
precast concrete
spandrel panel

Light framing and
cladding forms spandrel

▲ **9.5** Options for non-structural spandrel panels.

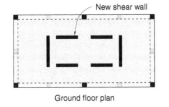

Soft storey
frame

Column
above

Column of new
two-way moment
frame

Ground floor plan

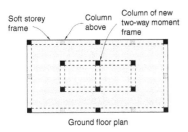

New shear wall

Ground floor plan

▲ **9.6** Interior moment frames or shear walls designed to resist all horizontal forces due to the unsuitability of perimeter soft storey frames.

you achieve this presumably architecturally desirable layout without creating a hazardous weak column–strong beam configuration? Initially, try applying the principle of separation. This means separating off the harmful excess strength of the beams from the frames in order to achieve a desirable weak beam–strong column moment frame. Remove the up- and down-stand spandrel elements often cast monolithically with beams. The beams become weaker than the columns and the moment frame becomes potentially ductile. A beam also becomes more flexible, so that aspect requires checking by the engineer. Perhaps the column dimensions will need enlargement to resist code-level seismic forces. Then clad the beams, now weaker than the columns, with spandrel panels. Any panel materiality may be chosen (Fig. 9.5). Panels are structurally separated from the moment frame to prevent them participating in force resistance and to avoid non-structural damage. Use one of the detailing approaches described in Chapter 11. In lieu of attached panels, spandrels can be fabricated from light framing and cladding attached directly to beams.

If, for any reason the strategy of separation is unacceptable, then consider differentiation as a solution. In this approach the seismically-flawed frame configuration remains on the façade but is relieved of any expectation of withstanding horizontal forces by provision of an alternative and stiffer system elsewhere in plan. The internal moment frames and shear walls of Fig. 9.6 resist *all* horizontal forces because they are stiffer and stronger. They are designed so as the perimeter frames need only carry gravity forces. The structural engineer might even intentionally soften-up columns of the seismically-flawed frames by introducing pins top and bottom to each column. This would prevent them attracting any horizontal forces at all. Whether or not that intentional softening is undertaken the perimeter frame must be flexible and possibly possess some ductility. It has to undergo the same horizontal drifts as the stronger alternative seismic resisting structure without distress while, at the same time, resisting gravity forces.

What if, as designer, you require a double-height floor at ground floor level, or anywhere else up the height of a building for that matter (Fig. 9.3(b))? Begin by accepting that the frames with such a flexible and soft storey must be excluded from the primary seismic resisting system. So keep their irregular configuration and design them to resist gravity forces only. Once again provide an alternative stiffer structure to resist *all* seismic forces (Fig. 9.6). The structural engineer will check that the soft storey frames can sustain anticipated horizontal drifts without damage.

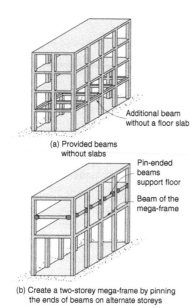

(a) Provided beams
without slabs

Additional beam
without a floor slab

Pin-ended
beams
support floor

Beam of the
mega-frame

(b) Create a two-storey mega-frame by pinning
the ends of beams on alternate storeys

▲ **9.7** Two methods of avoiding a soft storey where one storey is higher than others.

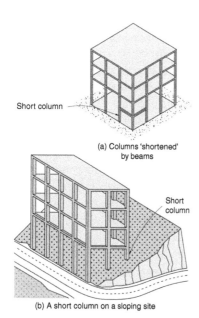

Short column

(a) Columns 'shortened'
by beams

Short
column

(b) A short column on a sloping site

▲ **9.8** Examples of short columns among longer columns of moment frames.

At least two other approaches are possible. First, introduce beams without floor slabs (Fig. 9.7(a)). This may achieve the intended spaciousness of the double-height storey yet avoid a soft storey by restoring the regularity of the moment frame. Now that the weight of the level without a floor is far less than that of the floor above, a special engineering analysis and design is required. If the idea of inserting beams to create regularity is unattractive, consider a mega-frame solution (Fig. 9.7(b)). The moment frame storey height is extended to two storeys. At alternate storeys floor beams are pinned at their ends to prevent them participating as moment frame elements. The main disadvantage of the mega-frame solution is that the frame member sizes are considerably larger than normal in order to control the increased drift and bending and shear stresses due to the increased storey heights. The columns must also be designed to resist mid-height inertia forces acting at alternate storeys.

Arnold and Reitherman suggest several other possibilities such as increasing the size and/or the number of columns in the soft storey or designing external buttresses to act as shear walls or braced frames for that storey.[3]

SHORT COLUMNS

Having just been informed of configuration difficulties posed by long columns, prepare to learn of the dangers of short columns. Structural extremes are unacceptable in seismic resistant systems. Short columns are to be avoided just as assiduously as flexible columns.

There are two types of short column problems; firstly, where some columns are shorter than others in a moment frame, and secondly, where columns are so short they are inherently brittle. The short columns of the second group are usually normal length columns that are prevented from flexing and undergoing horizontal drift over most of their height by partial-height infill walls or very deep spandrel beams.

Figure 9.8 shows examples where columns, some shorter than others in the same frame, cause seismic problems. The structural difficulty arising from these configurations is illustrated in Fig. 9.9. Two columns together, one that is half the height of the other, resist a horizontal force. The stiffness of a column against a horizontal force is extremely sensitive to its length; in fact, inversely proportional to the column length cubed (L^3). The shorter column is therefore eight times stiffer than the other, so it tries to resist almost eight times as much force as the longer column. It is unlikely to be strong enough to resist such a large proportion of the horizontal force and may fail.

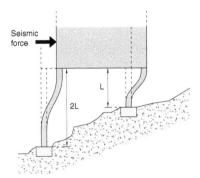

Seismic force

L

2L

▲ **9.9** Two unequal height columns resisting seismic force.

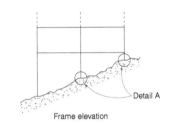

Detail A

Frame elevation

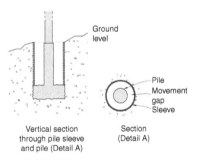

Ground level

Pile
Movement gap
Sleeve

Vertical section through pile sleeve and pile (Detail A)

Section (Detail A)

▲ **9.10** A method of avoiding a short column on a sloping site.

This type of short column problem can be overcome in several ways, some of which have already been described. An application of the differentiation approach will relieve the frame with one or more short columns of any responsibility for seismic resistance. Shear walls or braced frames are provided elsewhere in plan (Fig. 9.6). If the beams that frame into the columns forming short columns in Fig. 9.8(a) are pinned at both ends, that effectively doubles the column lengths and makes them all of equal length as far as seismic resistance is concerned. Of course, that creates a soft storey scenario that then needs to be addressed. An alternative approach to structuring Fig. 9.8(a) is to neglect the seismic strength of the long columns altogether and to resist all seismic forces by four one-bay frames; two acting in each direction to achieve a symmetrical structural configuration.

On a sloping site, short columns can be lengthened by integrating them with the piles (Fig. 9.10). If the piles are monolithic with columns and protected from contact with the ground by sleeves or casings that allow unrestrained horizontal drift, then a short column is avoided. Finally, check that a soft storey does not result from this foundation modification.

Now we return to short columns which have a very short distance over which they can flex horizontally (Fig. 9.11). As a rule-of-thumb a

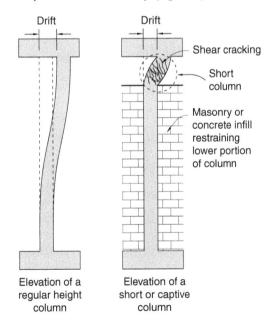

Drift

Drift

Shear cracking

Short column

Masonry or concrete infill restraining lower portion of column

Elevation of a regular height column

Elevation of a short or captive column

▲ **9.11** Comparison between a regular and a short column subject to horizontal drift.

▲ **9.12** Typical short column damage. 1994 Northridge, California earthquake.
(Reproduced with permission from Andrew B. King).

▲ **9.13** Short column failure. 2007 Pisco, Peru earthquake.
(Reproduced with permission from Darrin Bell).

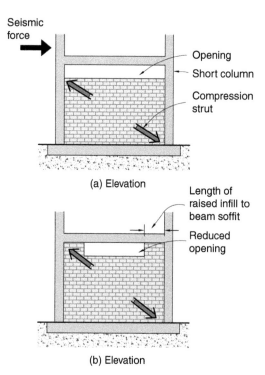

(a) Elevation

(b) Elevation

▲ **9.14** Reduction in the width of an opening above a partial-height masonry or concrete infill to prevent a short column failure. The raised lengths of infill enable a compression strut to transfer force directly to the top of the column and avoids the need for the column to bend.

short column has an unrestrained or free-length less that twice its depth. The problem is that the free-length is too short to allow for the development of a ductile plastic hinges. In the event of seismic overload the column fails in shear. To simulate a column snapping in a completely brittle manner, break a carrot between your hands. Once opposing diagonal shear cracks form in a reinforced concrete column its reduced gravity carrying capacity often leads to collapse (Figs 9.12 and 9.13).

Guevara and García describe this type of short column where its free-length is restricted by infill walls as a 'captive-column'. They explain where short columns are typically located in buildings and why they are so popular.[4] They also report on unsuccessful attempts by several international research groups to improve the seismic performance of short reinforced concrete columns, concluding that the best solution is to avoid them. If confined masonry or structural masonry walls (Chapter 5) are required to function as shear walls and the masonry is partial height, Guevara and García suggest continuing a short length of masonry up the sides of columns so that diagonal compression struts can act at the beam-column joint and thereby avoid short column failure (Fig. 9.14).

Chapter 10 discusses how non-structural partial-height masonry infills are separated to prevent short column

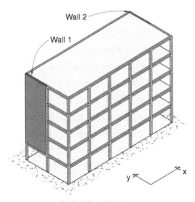

(a) Building with a discontinuous shear wall

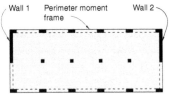

(b) Floor plan at an upper storey

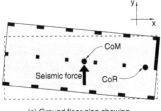

(c) Ground floor plan showing torsional rotation

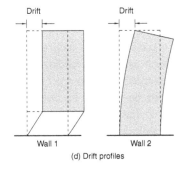

(d) Drift profiles

▲ **9.16** A discontinuous wall and its torsion-inducing influence on a building.

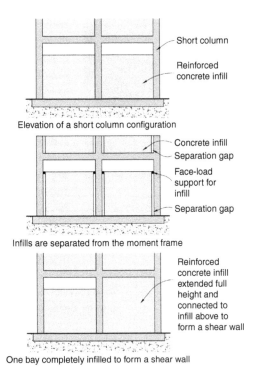

Elevation of a short column configuration

Infills are separated from the moment frame

One bay completely infilled to form a shear wall

▲ **9.15** Methods to avoid a short column configuration with reinforced concrete infills.

configuration. The same approach applies if infill walls are of reinforced concrete construction. Alternatively, designers can infill one or more windows to form shear walls which are strong and stiff enough to resist seismic forces without short columns being damaged (Fig. 9.15). Note that even if strong infills are separated from the moment frame as shown, the ductility of the frame is reduced due to the stiffening and strengthening effect the infills have on the beams. The beams cannot bend when the building sways, so large cracks form at column-beam interfaces. Some engineers specify thin horizontal slots, at least as long as the beam is deep, to be filled with soft material at each end of an infill. This detail avoids the extreme concentration of bending deformation at the ends of beams.

DISCONTINUOUS AND OFF-SET WALLS

Consider the building in Fig. 9.16. At its upper levels y direction forces are resisted by shear walls at each end, but at ground floor level the left-hand wall, Wall 1, is discontinuous. Two perimeter moment frames

▲ 9.17 Ground floor damage caused by a discontinuous wall. 1980 El Asnam, Algeria earthquake. (Bertero, V.V. Courtesy of the National Information Service for Earthquake Engineering, EERC, University of California, Berkeley).

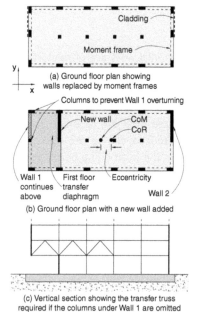

(a) Ground floor plan showing walls replaced by moment frames

(b) Ground floor plan with a new wall added

(c) Vertical section showing the transfer truss required if the columns under Wall 1 are omitted

▲ 9.18 Alternatives to a discontinuous wall.

resist *x* direction forces. When struck by a quake in the *y* direction, the ground pulses will distort the ground floor columns under Wall 1. Their 'softness' prevents Wall 1 from providing the seismic resistance perhaps expected of it and exemplifies the worst possible case of a soft storey. At the other end of the building the base of Wall 2, which is continuous, moves with the ground motion. Due to the more substantial overall strength and stiffness of Wall 2, as compared to Wall 1, Wall 2 tends to resist the inertia forces from the whole building. The two different wall drift profiles are shown in Fig. 9.16(d). Since Wall 1 resists almost no inertia force due to its discontinuity, yet Wall 2 is fully functional the building experiences serious torsion. To some degree, but limited by the modest lever-arm between them and their inherent flexibility, the two *x* direction moment frames try to resist the torsion. As the building twists about its CoR located close to Wall 2, the columns furthest away from the CoR are subject to large drifts and severe damage (Fig. 9.17).

What are the solutions to this problem? Probably the best option is to make both walls non-structural. Form them from either light-weight materials or use non-structural cladding panels to achieve the required architectural characteristics. Using the same approach as the building of Fig. 9.16, provide new moment frames behind the non-structural walls (Fig. 9.18(a)). Another possibility is to introduce an off-set single-storey wall back from Wall 1 (Fig. 9.18(b)). As explained below, this solution, which introduces many architectural and engineering complexities, is best avoided. Chapter 4 discusses this less-than-ideal situation where a transfer diaphragm channels seismic shear forces from the base of the upper section of a wall across to the top of an off-set wall below. This situation applies to Wall 1. Two strong columns, one at each end of Wall 1 must withstand vertical tension and compression forces to prevent it overturning under the influence of floor diaphragm forces feeding into it up its height. As mentioned in Chapter 4, if those columns are omitted, the overturning-induced axial force can also be resisted by two deep transfer trusses or beams. They must remain elastic during the design-level quake to prevent permanent downwards movement of the wall. As a rule-of-thumb, the truss depths should be between one and two times the cantilever span distance depending on the building height. In many cases such deep members, which must extend well into the body of the building to get sufficient vertical support to stabilize them, are not architecturally feasible (Fig. 9.18(c)). Since the trusses or deep beams create a strong beam–weak column configuration, ground floor shear walls in

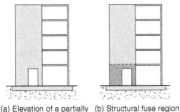

(a) Elevation of a partially discontinuous wall

(b) Structural fuse region at ground floor level

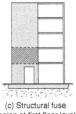

(c) Structural fuse region at first floor level

▲ **9.19** A partially discontinuous wall and options for the location of its structural fuse or plastic hinge region.

Inertia force

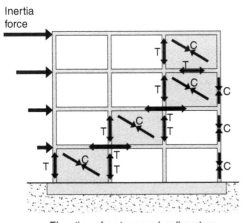

Elevation of a staggered wall system

▲ **9.20** The indirect force paths of a staggered-wall system.

the x direction will be required as well as the whole of the first floor slab being designed as a transfer diaphragm. Another reason the off-set solution is not ideal is that torsion is introduced due to eccentricity between the CoM and CoR at ground floor level for y direction forces.

The danger posed by off-set walls supported on cantilever beams has been tragically and repeatedly observed during five Turkish earthquakes in the 1990s. After categorizing building damage a report concludes: 'Buildings having architecturally based irregular structural systems were heavily damaged or collapsed during the earthquake. Cantilevers of irregular buildings have again proven to be the primary source/cause of seismic damage. Many buildings have regular structural systems but [even if] roughly designed performed well with minor damage'.[5]

Figure 9.19 shows a less extreme wall discontinuity. A large penetration weakens the most highly stressed region of the wall creating an undesirable soft storey. Traditional engineering wisdom would advise approaches as outlined previously, such as designing the wall to be non-structural, but given the sophisticated 3-D analysis and design tools available to contemporary structural engineers, a careful design might achieve satisfactory seismic performance. When approaching the design of an element with a discontinuity such as this, it is crucial that designers first identify the ductile overload mechanism (Fig. 9.19(b) or (c)), and then using the Capacity Design approach as described in Chapter 3, ensure dependable ductile behaviour. One approach is to design for plastic hinging at ground floor level and detail the wall and unattached column accordingly with the wall above strengthened to avoid premature damage. Another approach is for the first floor section of wall to be designated the fuse region. This means the ground floor section and the wall above first floor will be stronger than the fuse so damage occurs only in that specially detailed area.

In an extreme example of a staggered-wall (Fig. 9.20), the same principles apply. After computer analysis in order to examine the indirect force path, a fuse region must be identified and detailed to accept damage before any other structural element in the force path is affected. Due to its structural irregularity and complexity, as well as the difficulty of applying Capacity Design principles, all of which drive up the cost of construction, this system is not recommended.

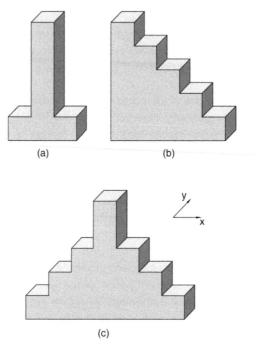

▲ **9.21** Typical setback configurations.

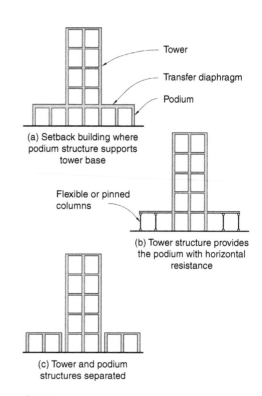

(a) Setback building where podium structure supports tower base

(b) Tower structure provides the podium with horizontal resistance

(c) Tower and podium structures separated

▲ **9.22** Different approaches to the configuration of a tower and podium building.

SETBACKS

A setback is where a plan dimension of a storey above a certain level in a multi-storey building reduces (Fig. 9.21). Seismic codes categorize buildings with abrupt setbacks as irregular. Sophisticated structural analyses quantify the 'notch effect' of a setback, but even though structural engineers avoid notches wherever possible because of stress concentrations, setbacks can be designed satisfactorily. The need for 3-D modelling of setback buildings can be appreciated from Fig. 9.21(b). Although the irregular vertical configuration in the *x* direction can be designed for, *y* direction shaking induces torsion due to the way the positions of the CoM and CoR change at every setback.

The podium or plaza and tower form represents a rather severe setback configuration (Fig. 9.21(a)). Designers are faced with several choices. They can treat the building as one structure. In this case, the podium roof is probably designed as a transfer diaphragm to force the podium framing to contribute to the horizontal force resistance at the bottom storey of the building (Fig. 9.22(a)). Alternatively, designers can provide

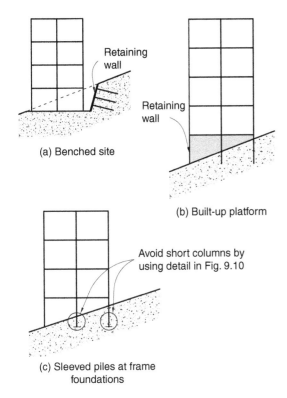

(a) Benched site

(b) Built-up platform

Retaining wall

Retaining wall

Avoid short columns by using detail in Fig. 9.10

(c) Sleeved piles at frame foundations

▲ **9.23** Structural configurations for moment frames on sloping sites.

the podium with little if any horizontal resistance and tie it strongly to the primary structure of the main tower, which then resists the seismic force of the entire building. Finally, the podium can be seismically separated from the tower. The tower then becomes a regular structure with more predictable seismic performance. Seismic separation joint treatment is discussed in the Chapter 8. A structurally independent podium must be seismically self-reliant so it requires its own seismic structure. The final choice should be made only after discussion between architect and structural engineer. Certainly a situation where the tower is off-set significantly in plan from the podium would encourage separation to avoid the additional complexity of torsion.

BUILDINGS ON SLOPING SITES

Building on a sloping site has already been raised with reference to avoiding short columns (Fig. 9.8(b)). The structural options for designers are now summarized in Fig. 9.23. In the first two options the ground is either benched and retained, or a strong and stiff built-up platform is formed by a walled retaining structure. Alternatively, the solution of Fig. 9.23(c) can be adopted to provide regular building configuration.

REFERENCES AND NOTES

1 For example, Structural Engineering Institute (2006). *Minimum Design Forces for Buildings and Other Structures (ASCE/SEI 7-05)*. American Institute of Civil Engineers. This code satisfies the seismic design requirements of the International Building Code 2006 (IBC 2006), published by International Code Council, Inc.

2 Comartin, C.D., Greene, M. and Tubbesing, S.K. (eds.) (1995). *The Hyogo-Ken Nanbu Earthquake: Great Hanshin Earthquake Disaster January 17, 1995, preliminary reconnaissance report*. Earthquake Engineering Research Institute.

3 Arnold, C. and Reitherman, R. (1982). *Building Configuration and Seismic Design*. John Wiley & Sons.

4 Guevara, T. and García, L. (2005). The captive- and short-column effects. *Earthquake Spectra*, **21**:1, 141–160.

5 Sesigür, H. et al (2001). Effect of structural irregularities and short columns on the seismic response of buildings during the last Turkey earthquakes. In Corz, A. and Brebbia, C.A. (2001). Earthquake resistant structures III. *Proceedings of the International Conference of Earthquake Resistant Engineering Structures (3rd)*. WIT, 83–90.

10 Non-structural elements: those likely to cause structural damage

Introduction

Imagine a reinforced concrete or steel building under construction. Visualize its structural framework, its beams, columns and floor slabs and its openness and emptiness. That is the building structure. Everything else yet to be constructed, all the remaining elements of construction and occupancy yet to be provided, fall into the category of non-structural elements.

Non-structural elements are, by definition, not intended to resist any seismic forces other than those resulting from their own mass. They are also, in the main, elements that structural engineers do not design and for which architects, and mechanical or electrical engineers take primary responsibility. The diverse types of non-structural elements can be divided into three groups:

- Architectural elements such as cladding panels, ceilings, glazing and partition walls
- Mechanical and electrical components like elevators, air conditioning equipment, boilers and plumbing, and
- Building contents, including bookcases, office equipment, refrigerators and everything else a building contains.

Non-structural elements, therefore, transform a structure into a habitable and functional building. The occupants, the fabric and contents of buildings and the activities undertaken in them are the life-blood of society. So it should come as no surprise that architects need to ensure, on behalf of their clients, that non-structural elements perform adequately during earthquakes. The two compelling reasons for taking the seismic performance of non-structural elements seriously are: firstly, the danger these elements pose to people both within and adjacent to the perimeter of a building, and secondly, the economic investment in buildings and enterprises occurring within them.

The value of non-structural elements expressed as a percentage of the total cost of a building, excluding the price of the land, depends upon the type of building considered. In an industrial or storage building with few mechanical services and architecturally designed elements, non-structural costs can be in the 20 to 30 per cent range. In more heavily serviced and complex buildings non-structural elements can comprise up to 85 per cent of the total cost. In some cases, such as an art museum or a high-tech research or computer centre, the value of non-structural elements, especially the building contents, may well exceed the sum of all other costs associated with the building. It makes sense to protect this investment from earthquake damage. Non-structural elements have proven to be very vulnerable to seismic damage as evidenced by the 1994 Northridge, California earthquake. If a quarter of the most severely damaged of the 66,000 buildings surveyed are excluded, most of the remaining damage occurred to non-structural elements.[1]

In a seismic study of a 27-storey building in Los Angeles subject to the Maximum Credible Event, direct economic loss of non-structural elements exceeded by six times the cost of structural damage.[2] This analysis excluded the cost of indirect losses of revenue and building use. For some buildings these less tangible but nonetheless real costs will greatly exceed the value of direct losses. While emergency services, hospitals and similar facilities need to be operational immediately after a damaging quake, more and more businesses are becoming aware of how important business continuance is to their financial viability in a post-earthquake environment.

During earthquake shaking, non-structural elements represent a significant hazard to people. Injuries are caused by building elements such as glazing or suspended ceilings shattering or collapsing, or by building contents being flung around. Filing cabinets and equipment overturn, containers of hazardous materials break open or gas from ruptured pipes ignite. Damage scenarios vary from building to building and room to room. Try to image what damage might occur if the room you are in now is suddenly shaken and its fabric such as partitions, windows and ceiling are damaged, and contents, including yourself, are flung about. However, it is very likely your risk of injury is far less than if you were in other more hazardous locations; like walking down a supermarket

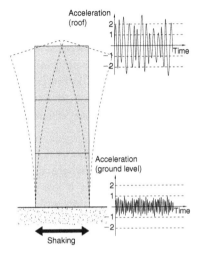

▲ 10.1 Shaking at ground level is amplified up the height of a building.

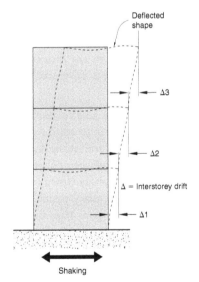

▲ 10.2 Interstorey drifts from earthquake shaking.

aisle between high storage racks or against a building with weakly attached cladding panels.

So what is the root cause of non-structural damage? Although the obvious reason is that a building is shaken by an earthquake, it is worth singling out two specific aspects of seismic shaking that explain observed damage; acceleration and interstorey drift. As explained in Chapter 2 and illustrated in Fig. 10.1, a building amplifies ground accelerations as it resonates in response to the dynamic movement at its base; the higher-up a building the greater the level of acceleration amplification. Horizontal accelerations induce inertia forces in non-structural elements that cause them to slide, overturn, break loose from their attachments to the main structure and both suffer and cause further damage. The two strategies for preventing such damage is: firstly, to ensure that non-structural elements themselves have enough strength to resist their own inertia forces; and secondly, to physically restrain the elements by attaching them to structural members. Chapter 11 considers these requirements in detail.

The second type of non-structural damage is caused by interstorey drift. As a building sways to-and-fro during a quake each floor drifts or deflects horizontally further than the floor below (Fig. 10.2). Interstorey drift or relative horizontal movement between floors can damage non-structural elements that connect to both floors. As discussed below, careful separation of non-structural elements from the structure avoids this damage.

While the damage scenarios above identify some of the safety hazards people face in the immediate presence of non-structural elements, this chapter also explores a less obvious but more serious situation. That is, the likelihood of non-structural elements damaging the primary structure of a building to such an extent as to cause partial collapse with ensuing risk of injury and huge economic losses. The two types of non-structural elements most capable of causing global rather than local building damage, and may also be hazardous for people in their vicinity, are infill walls and staircases.

INFILL WALLS

Infill walls are non-structural walls constructed between columns. Where located on the exterior of a building as part of the cladding system, infill walls usually are bounded by structure; columns on either side, floor surfaces below and beams above. A beam may not necessarily be present but most infill walls abut columns. The description of most infill walls as 'non-structural' is misleading to say the least.

Although they are not designed to resist either gravity or horizontal forces no one has informed *them*! By virtue of their inherent in-plane strength and stiffness they cannot avoid resisting forces even if they wanted to. *Any* stiff and strong building elements, whether designed by structural engineers or not, attract forces to themselves. In the process of resisting seismic forces, infill walls can cause serious structural damage to a building. That is why the problems they cause, and solutions to overcome them, require careful consideration.

As discussed in Chapter 5 infill walls can helpfully resist seismic forces in buildings, but only in certain situations. These include where there is no other seismic resisting system provided; the building is low-rise; the masonry panels are continuous from foundation to roof; there are enough panels in each plan orthogonal direction to adequately brace the building; the infills are not heavily penetrated; and finally, where infill walls are placed reasonably symmetrically in plan. Most infill walls do not satisfy these criteria and may introduce configuration deficiencies (Chapter 9).

Particular care is required when adding or modifying infill walls during building alterations. If infills, including those that can be categorized as confined masonry walls (Chapter 5), are to function as shear walls they should not be penetrated nor have existing openings enlarged without engineering advice. A similar cautionary note applies to any insertion of infill walls that might detrimentally affect the seismic performance of the building by, for example, causing torsion.

Infill walls that are capable of causing structural damage to a reinforced concrete or structural steel frame building are usually constructed from solid or hollow masonry bricks or concrete blocks that are usually unreinforced and plastered. A large concrete panel placed between columns also constitutes an infill. In wood construction, gypsum plasterboard infill walls are also strong and rigid enough to disrupt the primary structure. Infill walls are usually constructed after columns and beams have been cast or erected. They are stiff, strong and brittle when forced parallel to their lengths (in-plane) yet vulnerable to out-of-plane forces.

Problems associated with infill walls

So what are the difficulties with infill walls given that they are commonly used in so many countries? Why do they require special attention in seismically active regions?

Firstly, infill walls stiffen a building against horizontal forces. As explained in Chapter 2, additional stiffness reduces the natural period of vibration, which in turn leads to increased accelerations and inertia forces (Fig. 10.3). As the level of seismic force increases, the greater the likelihood

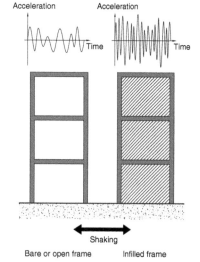

▲ **10.3** A comparison of roof-top accelerations of a bare or open frame with an infilled frame. Note the shorter periods of vibration and higher accelerations of the infill frame.

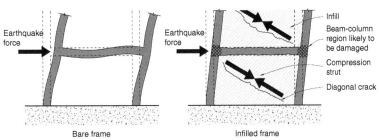

▲ **10.4** Whereas a bare frame deflects horizontally by columns and beams bending, the stiffness of a masonry infill limits horizontal movement. A diagonal compression strut forms together with a diagonal tension crack caused by elongation along the other diagonal.

▲ **10.5** Typical infill wall diagonal crack pattern. 1999 Chi-chi, Taiwan earthquake. (Reproduced with permission from Geoff Sidwell).

▲ **10.6** Damage to the tops of several columns due to infill wall compressive strut action. Mexico City, 1985 Mexico earthquake. (Reproduced with permission from R.B. Shephard).

of non-structural as well as structural damage. To some degree, the force increase can be compensated for by the strength of the infills provided they are correctly designed to function as structural elements.

Secondly, an infill wall prevents a structural frame from freely deflecting sideways. In the process the infill suffers damage and may damage the surrounding frame. The in-plane stiffness of a masonry infill wall is usually far greater than that of its surrounding moment frame – by up to five to ten times! Without infill walls a bare frame deflects under horizontal forces by bending in its columns and beams. However, a masonry infill dominates the structural behaviour (Fig. 10.4). Rather than seismic forces being resisted by frame members, a diagonal compression strut forms within the plane of the infill, effectively transforming it into a compression bracing member. Simultaneously, a parallel diagonal tension crack opens up between the same two corners of the frame because of the tensile elongation along the opposite diagonal and the low tensile strength of the infill material. The infill panel geometry deforms into a parallelogram. After reversed cycles of earthquake force, 'X' pattern cracking occurs (Fig. 10.5). The strength of the compression strut and the intensity of force it attracts concentrates forces at the junction of frame members. Shear failure may occur at the top of a column just under the beam soffit (Fig. 10.6). Such a failure is brittle and leads to partial building collapse.

During a damaging quake diagonal cracks and others, including those along the interface of infill and columns and the beam above, soften-up the infill. It becomes weaker and more flexible than a less severely damaged infill above it – in effect creating a soft-storey (Chapter 9). Even if infill walls are continuous vertically from the foundations to roof, once ground floor infill walls are damaged a soft storey failure is possible.

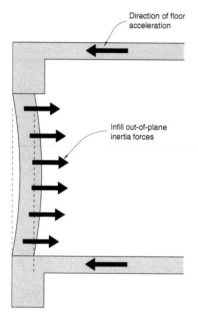

▲ 10.7 A section through two floors and an infill wall. Out-of-plane forces act on the infill which spans vertically between floors.

Another danger facing a heavily cracked infill is its increased vulnerability to out-of-plane forces (Fig. 10.7). The wall may become disconnected from surrounding structural members and collapse under out-of-plane forces. Due to their weight, infill walls pose a potential hazard to people unless intentionally and adequately restrained.

The final problem associated with the seismic performance and influence of infill walls is that of torsion (Chapter 2). Unless infill walls are symmetrically placed in plan their high stiffness against seismic force changes the location of the Centre of Resistance (CoR). In Fig. 10.8(a) the CoR and Centre of Mass (CoM) are coincident; no significant torsion occurs. If infill walls are located as in Fig. 10.8(b), the CoR moves to the right and the subsequent large torsional eccentricity causes the building to twist when forced along the y axis (Fig. 10.8(c)). As one floor twists about the CoR relative to the floor beneath the columns furthest away from the CoR sustain large interstorey drifts and damage. If the drifts are too large, those columns are unable to continue to support their gravity forces and their damage leads to that area of the building collapsing. In this example, the infill walls cause torsion during y direction shaking only.

Solutions to problems caused by infill walls

Unfortunately, only three solutions are available: the first is often not feasible and the other two, while simple in theory, are difficult to achieve in practice.

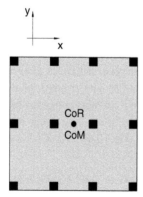

(a) Symmetrical frame building with CoR and CoM co-incident

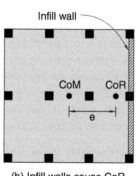

(b) Infill walls cause CoR to move to right creating an eccentricity between CoM and CoR

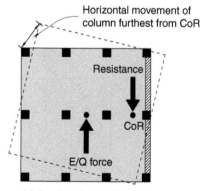

(c) A y direction earthquake force causes the building to twist about CoR

▲ 10.8 Asymmetrically placed infill walls cause building torsion that damages columns distant from the CoR.

The first solution (as discussed in Chapter 5) is for infill walls to be transformed into confined masonry construction that is fully integrated with the structural frame. As discussed in Chapter 5, confined masonry can play an important role in providing seismic resistance. However, this approach is valid only where designs comply with the numerous and architecturally restrictive criteria that apply to confined masonry construction.

Another alternative to overcoming the problems associated with infill walls is to provide very stiff primary structure. In this situation the less stiff infill walls do not attract horizontal forces. Reinforced concrete (RC) shear walls are the only structural system capable of achieving the necessary stiffness. Only such stiff structure can limit interstorey drifts to several millimetres per floor during seismic shaking. Commenting from a European perspective, Michael Fardis suggests that: 'The best way to protect an RC building from the adverse effects of heightwise irregular infilling is by providing shear walls that are strong and stiff enough to overshadow any difference between the infilling of different storeys.'[3] Later he points out that the European seismic code takes a more permissive approach.[4] It allows masonry infills to be present in frame structures provided certain rules are followed. For example, if the plan layout of infills is asymmetrical, possible adverse torsional effects must be investigated by extensive 3-D computer modelling. Where infills are distributed vertically in an irregular pattern causing a potential soft storey, affected columns are required to be far stronger and larger than usual. Where such columns are to be designed, and assuming they remain elastic in the design-level quake, their design forces may be of the order of five times greater than those of ductile columns.

This leaves the third and final option able to be summarized in a single word – *separation*. Based upon the relatively poor seismic performance of infill walls in past earthquakes, current practice in seismically active countries such as Japan, USA and New Zealand is to separate infill walls from their frames. Where a country's seismic design philosophy requires that non-structural elements escape damage in small earthquakes, and do not damage primary structure in a large event, separation becomes the most common solution. Separation gaps allow the frame to deflect freely without being impeded by the wall (Fig. 10.9).

Infill walls require separation from the frame by gaps of sufficient width as calculated by the structural engineer. Separation gaps provide architectural detailing challenges. Issues such as acoustic control, weather tightness, fire protection and aesthetic qualities all need to be addressed.

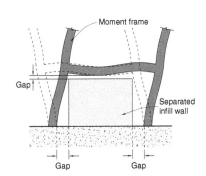

▲ **10.9** Infill wall with separation gaps between infill and columns and beam.

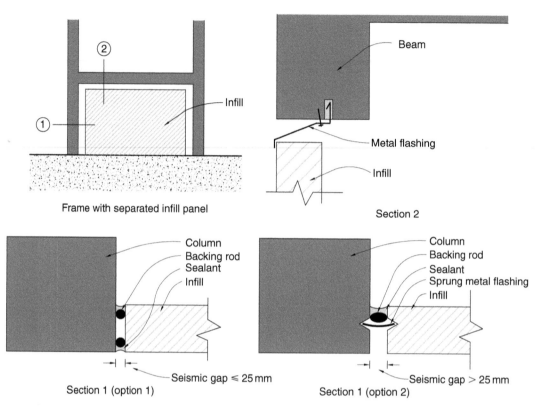

Frame with separated infill panel

Section 2

Section 1 (option 1)

Section 1 (option 2)

▲ 10.10   Some typical architectural details of separation gaps between an infill wall and frame.

Resolution of these architectural details is commonplace in the countries listed above. Several typical details are shown in Fig. 10.10.

Two essential features of a seismically separated infill wall are: a clear vertical gap between the infill and columns (typically between 20 mm to 80 mm wide), and an approximately 25 mm wide horizontal gap between the top of the wall and the soffit of the beam above. This gap under a beam or floor slab must be greater than that element's expected long-term deflection, and also allow for the downwards bending deflection of a moment frame beam under seismic forces. Where provided, these gaps allow the floor above an infill to move horizontally to-and-fro without the infill wall offering any resistance in its plane.

Often the resolution of one problem creates another. Although an infill may be separated for *in-plane* movement, where it is separated on three sides it becomes extremely vulnerable to *out-of-plane* forces or *face-loads* as they are sometimes called. It must be stabilized against

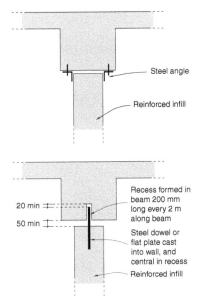

Steel angle

Reinforced infill

20 min

50 min

Recess formed in beam 200 mm long every 2 m along beam

Steel dowel or flat plate cast into wall, and central in recess

Reinforced infill

▲ **10.11** Two possible structural details that resist out-of-plane forces yet allow relative movement between an infill wall and structure above.

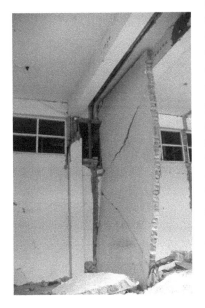

▲ **10.13** Partial out-of-plane collapse of an unreinforced masonry infill wall. Commercial building, Tarutung, Indonesia, 1987 Sumatra earthquake.

▲ **10.12** A reinforced masonry infill wall separated from surrounding structure. Note the horizontal and vertical gap (to beam) and a galvanized steel bracket resisting out-of-plane forces yet allowing in-plane movement. Office building, Wellington.

these forces acting in its weaker direction, yet at the same time allow unrestricted inter-storey drift along its length. One of several structural solutions is required.

The most obvious approach to stabilizing an infill wall against out-of-plane forces is to cantilever it from its base. But this is not usually feasible for two reasons. Firstly, the floor structure beneath may not be strong enough to resist the bending moments from the wall. Secondly, the infill wall itself may not be strong enough or may require excessive vertical reinforcing. On the upper floors of buildings, elements like infill walls are subject to very high horizontal accelerations well in excess of 1 g, the acceleration due to gravity.

The preferred option is to design an infill wall to resist out-of-plane forces by spanning vertically between floors (Fig. 10.7). Through its own strength the wall transfers half of its inertia force to the floor beneath and the other half to structure above. Careful structural detailing at the top of the wall can provide sufficient strength to prevent out-of-plane collapse yet simultaneously accommodate interstorey drift between the top of the wall and structure above. Figure 10.11 illustrates some generic connections between reinforced masonry infill walls and concrete frames while Fig. 10.12 illustrates an as-built solution.

Where infill walls are constructed from unreinforced masonry – which is generally too weak to span vertically from floor to floor when withstanding out-of-plane inertia forces (Fig. 10.13) – one approach is to provide small reinforced concrete columns within the wall thickness (Fig. 10.14). Their function is not to support any vertical force but to stabilize the infill against out-of-plane forces. For a long panel, three or more intermediate 'practical columns' (as they are sometimes called)

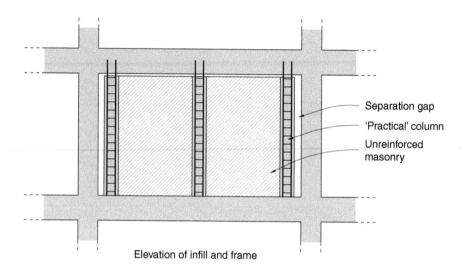

Elevation of infill and frame

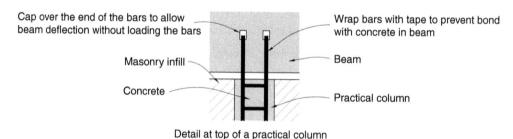

Cap over the end of the bars to allow beam deflection without loading the bars

Wrap bars with tape to prevent bond with concrete in beam

Masonry infill

Beam

Concrete

Practical column

Detail at top of a practical column

▲ **10.14** Separated unreinforced masonry infill wall with 'practical columns' providing out-of-plane strength.

may be designed by the structural engineer.[5] Support of this form is commonplace near the roof of a building where ground accelerations are amplified most strongly. Only the reinforcing bars that project vertically from these small columns connects to the underside to the beam. This detail, strong enough to resist out-of-plane forces, allows virtually unrestrained inter-storey drift in the plane of the wall. The ductile bending of the vertical practical columns bars will not provide significant resistance to that movement.

It is worth noting that some of the above separation difficulties can be alleviated by off-setting intended infill walls from the primary columns. Walls are therefore designed to run in front of or behind columns. No longer infills, they can be considered partition walls or exterior cladding and are discussed in Chapter 11.

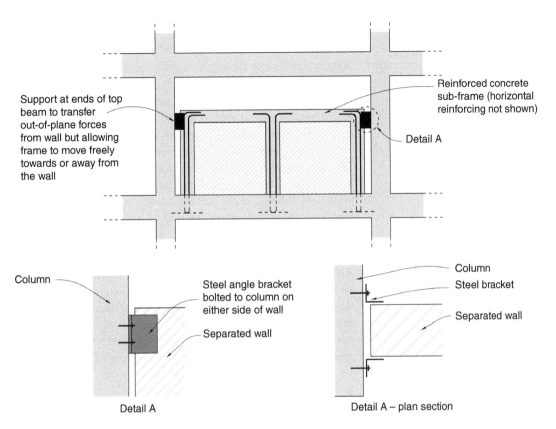

Support at ends of top beam to transfer out-of-plane forces from wall but allowing frame to move freely towards or away from the wall

Reinforced concrete sub-frame (horizontal reinforcing not shown)

Detail A

Column

Steel angle bracket bolted to column on either side of wall

Separated wall

Detail A

Column

Steel bracket

Separated wall

Detail A – plan section

▲ **10.15** The separation of a partial-height unreinforced masonry infill wall.

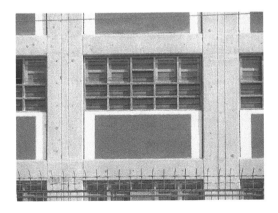

▲ **10.16** A separated partial-height infill in a moment frame. Due to the low height of the infill its out-of-plane inertia forces can be resisted by two small reinforced concrete columns. Office building, Pisco, Peru. (Reproduced with permission from Darrin Bell).

Partial-height infills also need separation from their adjacent columns yet be prevented from collapsing out-of-plane. A partial-height unreinforced infill wall can be enclosed within a reinforced concrete sub-frame that is then restrained out-of-plane by a steel bracket at each end or by some other structural solution (Figs 10.15 and 10.16). A clear gap must exist between the sub-frame and the primary structural frame.

An alterative detail is to provide a horizontal member, perhaps in the form of a channel section, to span between the columns of the main frame. The channel resists the upper half of the infill wall inertia forces and transfers them back to the main columns. The infill is free to slide

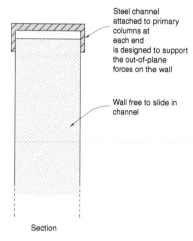

Steel channel attached to primary columns at each end is designed to support the out-of-plane forces on the wall

Wall free to slide in channel

Section

▲ **10.17** Alternative support to a partial-height infill wall. Depending on the wall height and the distance between primary columns the steel channel might require strengthening with welded-on plates.

in the direction of the channel and is of course separated by vertical gaps adjacent to the main columns (Fig. 10.17).

Another promising approach to the problem of infills currently under development is to 'soften' infills rather than fully separating them from frames.[6] This entails using weak mortar between bricks and laying bricks between vertical studs and horizontal dwangs to subdivide walls into many small areas. Out-of-plane support is provided by the studs. In a damaging quake these infills respond by 'working' along the joints between the infilling and frame members. The sliding between masonry and framing dissipates a significant amount of seismic energy. Relatively weak and flexible, these softened infills are sacrificial elements that provide increased levels of damping to protect the primary structure. They may be a useful strategy to improve the seismic resistance of new and existing moment frames, particularly in developing countries.

Although the need for infill separations is usually seen as a problem to be overcome, it can also lead to innovations. In Fig. 10.18 the insertion of narrow and tall windows between column and wall suggests another possible response. Provided that clearances around the glazing can accommodate calculated inter-storey drifts this detail avoids the need for a specific separation gap between the infill wall and columns.

STAIRCASES

Like infill walls, staircases damage the primary structure as well as being damaged by it. Where staircases are strongly attached to the structure, to some degree they act as structural members. Their inclination creates the potential for them to function as diagonal braces (Fig. 10.19).

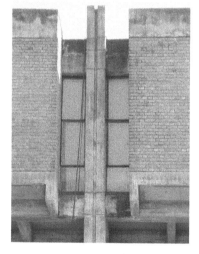

▲ **10.18** Full-height windows adjacent to columns suggest another means of creating vertical gaps between infill walls and columns. Library building, Kanpur, India.

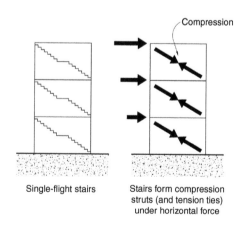

Compression

Single-flight stairs | Stairs form compression struts (and tension ties) under horizontal force

▲ **10.19** Bracing action of stairs connected to structure.

▲ **10.20** Damage to stair support structure. Mexico City, 1985 Mexico earthquake.
(Reproduced with permission from R.B. Shephard).

Braces that form a triangulated framework are very stiff against horizontal forces as compared to moment frames. Therefore, staircases can attract unanticipated high levels of force (Fig. 10.20). If stairs are severely damaged, building occupants may be unable to exit a building after an earthquake.

Where stairs are positioned asymmetrically in plan they induce building torsion. With reference to Fig. 10.8 assume a staircase is located adjacent to the right-hand edge of the floor plan. Before the staircase is constructed the building is completely symmetrical from a structural perspective, but if we assume that the staircase stiffness is equivalent to that of two infill walls, it moves the CoR to the right, creating torsion as explained previously.

To avoid damage to both staircase and structure, the recommended solution is to separate the stairs by providing a sliding joint at each floor (Fig. 10.21). The sliding detail must allow inter-storey drift between floors in any direction without restraint. When inter-storey drifts do occur, the stairs slide on the floors below and therefore do not attract any seismic force apart from the minimal inertia forces arising from their relatively small self-weight.

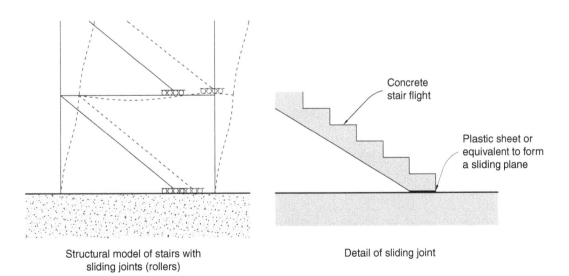

Structural model of stairs with
sliding joints (rollers)

Concrete
stair flight

Plastic sheet or
equivalent to form
a sliding plane

Detail of sliding joint

▲ **10.21** Model and detail of a single-flight stair separation.

Sliding joints which structurally separate stairs from primary structure are formed easily. Just break any likely bond between the stair and its base support; when the floor at the top of the stair drifts further than the floor below, the stair slides. The stair can be either pin jointed or rigidly cast into the floor above. Remember that all stair inertia forces, including forces at right angles to the direction of the staircase, must be transferred back to the main structure. Some stairs are separated with more sophisticated materials such as Teflon strips that bear on stainless steel plates. This combination possesses very low friction but is usually unnecessarily sophisticated for such a simple slip joint. Figure 10.22 illustrates movement detailing at the base of a staircase that an architect has chosen to celebrate.

The separation details for a switch-back or dog-leg stair with a half storey-height landing is more complex, as shown in Fig. 10.23.

▲ **10.22** A staircase celebrating provision of interstorey drift through rollers at its base. Office building, Wellington.

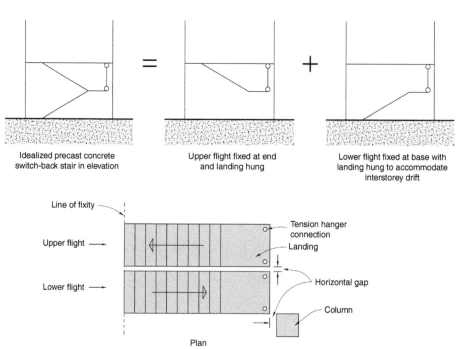

Idealized precast concrete switch-back stair in elevation

Upper flight fixed at end and landing hung

Lower flight fixed at base with landing hung to accommodate interstorey drift

Line of fixity

Tension hanger connection

Landing

Upper flight

Horizontal gap

Lower flight

Column

Plan

▲ **10.23** Idealized elevations and plan showing a method of separating a switch-back stair from the structure.

REFERENCES

1 MacLeod, F. (2004). The latest developments in seismic mitigation of suspended ceiling systems. *Proceedings of the 13th World Conference on Earthquake Engineering*, Vancouver, Paper No. 3364, 13 pp.

2 Arnold, C. (1991). The seismic response of non-structural elements in buildings. *Bulletin of the New Zealand National Society for Earthquake Engineering*, **24**:4, 306–316.

3 Fardis, M. (2006). Seismic design issues for masonry-infilled RC frames. *First European Conference on Earthquake Engineering and Seismology, Geneva, Switzerland*, Paper no. 312, 10 pp.

4 CEN (2004). *European Standard EN 1998-1:2004 Eurocode 8: Design of Structures for Earthquake Resistance. Part 1: General rules, seismic actions and rules for buildings.* Comite Europeen de Normalisation.

5 Hollings, J.P. et al. (1981). *Indonesian Earthquake Study: Vol. 6, Manual for the design of normal reinforced concrete and reinforced masonry structures.* Beca Carter Hollings and Ferner Ltd., Wellington (unpublished).

6 Langenbach, R. et al. (2006). Armature Crosswalls: A proposed methodology to improve the seismic performance of non-ductile reinforced concrete infill frame structures. *Proceedings of the 8th US National Conference on Earthquake Engineering (8NCEE)*, San Francisco. Paper no. 523, 10 pp.

11 OTHER NON-STRUCTURAL ELEMENTS

The previous chapter began by overviewing the diverse range of non-structural elements that distinguish a bare unfinished structure from a fully operational building. It then explained how non-structural elements need careful seismic design to prevent both injury and economic loss during earthquakes. After noting the two primary reasons for non-structural damage, namely seismic acceleration and interstorey drift, the chapter discussed infill walls and staircases, both of which can inflict serious damage upon the primary structure of buildings.

This chapter examines the remaining non-structural elements. While they do not pose a threat to structural elements, they do need to be either tied back to structure or separated from it for the sake of injury prevention and economic sustainability, including losses arising from a damaged building not being functional. A large proportion of people requiring medical treatment in emergency facilities after an earthquake are injured by moving or falling non-structural elements. In moderate earthquakes non-structural damage is the largest contributor to overall earthquake damage costs.

In spite of the importance of non-structural elements to the satisfactory day-to-day functioning of a building, their seismic performance is often not a high priority of building codes or the building industry in general. Some seismic codes require reduction of risk of death and injury from structural and non-structural elements but do not place importance on protecting the fabric and contents of buildings from damage. Architects need to be aware of limitations in their current seismic codes' provisions in this regard and, even if code requirements are satisfied, they should inform their clients of the expected damage that non-structural elements are likely to sustain during small and large earthquakes. On the basis of that information clients might request more resilient non-structural elements less susceptible to earthquake damage.

Although a code might require seismic protection to non-structural elements, this may not be achieved in practice unless a structural engineer is specifically involved. Many practical and organizational difficulties conspire to reduce the seismic safety and performance of non-structural elements.[1,2] These can be overcome only by a change in culture within design and construction teams including detailed attention to the seismic provisions for non-structural elements at the design, documentation and construction phases of a project. A higher standard of seismic performance requires additional professional input. Ideally, a member of the design team − such as the structural engineer − should be commissioned to take responsibility for the design and implementation of code seismic requirements for non-structural elements.

On that rather cautionary note, each type of non-structural element is considered in turn.

CLADDING

The term cladding refers to the non-structural external walls or skin of a building. From an international perspective claddings comprise a wide range of materials that reflect the diversity of built environments. For example, external walls of light-weight woven bamboo matting are found in areas of Asia and sun-dried adobe blocks in South America. Unreinforced masonry, which may or may not be plastered, is prevalent in almost every country. In more developed countries claddings such as glass fibre reinforced cement (GFRC) panels, titanium sheets and fully-glazed curtain walls are observed. Three categories of claddings: masonry, panels and other materials are considered in the following sections.

Masonry

Seismic considerations for masonry cladding are similar but less complex than those for infill walls (Chapter 10). Although external masonry walls may not be placed between columns and so by definition are not infill walls, they share some of the same undesirable seismic characteristics. Therefore, unless very stiff shear walls resist horizontal forces in the direction of their lengths, non-structural masonry walls should be separated from the main structure. The strategy of separation also

▲ **11.1** Unseparated masonry cladding damaged by the interstorey drift of a flexible internal steel frame. Mexico City, 1985 Mexico earthquake.
(Reproduced with permission from R.B. Shephard).

▲ **11.2** Collapse of exterior wall. 1983 Coalinga, California earthquake
(Reproduced with permission from R.B. Shephard).

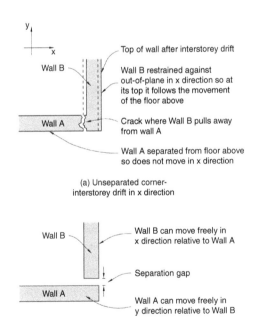

(a) Unseparated corner-
interstorey drift in x direction

(b) Separated corner detail

▲ **11.3** Plan of two walls forming an exterior corner. Because both walls need out-of-plane support where they connect to the structure above, interstorey drift damages the corner (a) unless a vertical separation gap, yet to be treated architecturally, is provided (b).

prevents indirect damage to the primary structure by eliminating potential configuration problems (Chapter 9) as well as in-plane cracking and more severe damage to the walls (Figs 11.1 and 11.2). Details of out-of-plane restraints like those of Fig. 10.11 are appropriate. A vertical separation gap at wall corners is required to control damage in those areas (Fig. 11.3).

Brick veneer is also a very popular cladding system. Unfortunately, it also doesn't have a good earthquake track record.[3] As an example, considerable the veneer damage that occurred during the 1989 Newcastle, Australia, earthquake (Fig. 11.4). Veneers support their own weight but, as their name implies, a veneer must be tied back to internal structure. Reinforced masonry walls or vertical wood or steel studs provide out-of-plane resistance by transferring inertia forces from the veneer to floor and ceiling diaphragms (Fig. 11.5). In wooden buildings, veneer ties are embedded in horizontal mortar joints and screwed to vertical posts or studs.

The two most important design principles for achieving the best seismic performance from a veneer are:

- Tie the veneer strongly to the structure for both tension and compression forces
- Veneer tie spacing should comply with code requirements. In New Zealand, ties are typically placed no further apart than 600 mm horizontally and 400 mm vertically,[4] and

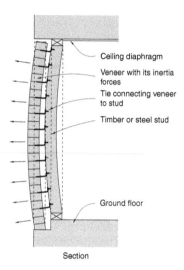

Ceiling diaphragm

Veneer with its inertia forces

Tie connecting veneer to stud

Timber or steel stud

Ground floor

Section

▲ **11.5** Inertia forces acting on a veneer are transferred through ties to studs and then to diaphragms above and below.

▲ **11.4** Damage to brick veneer due to corroded veneer ties. 1989 Newcastle, Australia earthquake.
(Reproduced with permission from R.B. Shephard).

- Use veneer ties that are flexible in the direction of the plane of the veneer itself unless the primary structure is at least as stiff as the veneer in that direction against horizontal forces. This allows the primary seismic resisting structure to deflect horizontally without loading and damaging the often stiffer veneer.

Relative flexibility between veneer and structural framework leads to a concentration of damage where veneer panels meet at corners. Damage prevention, necessitating wide vertical separation gaps at corners, is usually deemed impractical or aesthetically unacceptable for most buildings and so damage in those areas is often accepted as inevitable.

Due to its hazard if it were to fall from a building, the maximum height of veneer panels is limited by some codes. Where brick veneer is used for cladding multi-storey buildings, steel angles or other means of supporting the weight of the veneer are provided at each storey (Fig. 11.6). Increased earthquake accelerations up the height of a building might necessitate reduced tie-spacing and perhaps special horizontal reinforcement for additional safety.

▲ **11.6** Veneer on a multi-storey wood framed building. Each storey-height of veneer is supported on a steel angle bolted to framing. Hotel, Tongariro National Park, New Zealand.

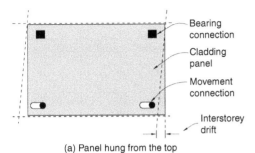

Bearing connection

Cladding panel

Movement connection

Interstorey drift

(a) Panel hung from the top

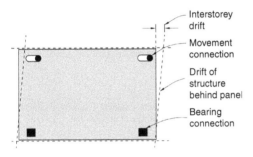

Interstorey drift

Movement connection

Drift of structure behind panel

Bearing connection

(b) Panel supported at its base

▲ **11.7** Fully separated storey-height panel, top-hung (a) and supported at its base (b).

▲ **11.9** Slotted steel connection welded to a plate cast into the column (a) and the completed detail allowing movement (b).

▲ **11.8** Precast panels attached to a reinforced concrete frame building. Note the connections near the tops of columns and the plane of horizontal separation between panels below the beam soffit. Office building, Wellington.

Panels

Although cladding panels are sometimes fabricated from relatively light-weight materials like fibreglass, most are concrete. They represent a serious hazard should they fall from a building. They can also damage the primary structure if their fixings prevent interstorey drift. For these reasons designers separate panels from the structure and from one another unless the panels are strongly connected to the structure and designed to act as shear walls (see Fig. 5.12). Just as snake and fish scales remain attached to the body yet flex relative to each other when movement occurs, so must cladding panels allow a primary structural frame to move to-and-fro without offering any resistance. Fig. 11.7 shows two possibilities of connecting yet separating storey-height panels. One method is to hang a panel from the top and create a clear horizontal gap between it and the panel beneath. Alternatively, a panel is supported at its base with the gap to accommodate interstorey drift located at the top of the panel. In both cases, even though the structural frame behind undergoes interstorey drifts, the panels remain vertical and each moves with the floor that provides its gravity support. All four panel connections resist out-of-plane forces. Two on one level allow horizontal seismic movement and the others need to accommodate only small contractions and elongations from shrinkage and temperature variations. A common detail that allows movement yet resists out-of-plane forces consists of bolts passing through slotted holes in steelwork connected to the structure (Figs 11.8 and 11.9). Given that panels should never

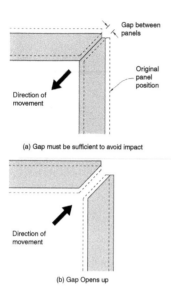

(a) Gap must be sufficient to avoid impact

(b) Gap Opens up

▲ **11.10** Provision for horizontal movement where panels meet at a corner as shown in plan.

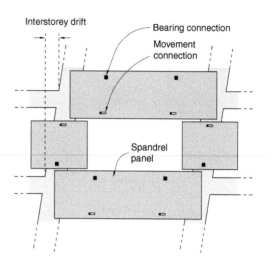

▲ **11.11** One approach to isolating spandrel and column panels from the interstorey drifts of a structural frame.

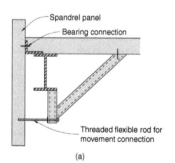

(a)

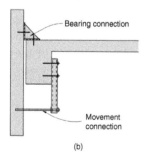

(b)

▲ **11.12** Spandrel panel bearing and movement connection details for a steel frame (a) and a reinforced concrete frame (b).

fall from a building, even during a severe quake, slot lengths are calculated by the structural engineer and details are designed to be as ductile as possible. Attention also needs to be paid to panel movement at external corners (Fig. 11.10).

Many variations in panel arrangements and connections are possible for spandrel panels and column cover panels (Fig. 11.11). Detailing must accommodate the interstorey drift. Spandrel panels are usually attached to the structural frame by bearing connections that support the weight of the panel at the middle to top of the panels. Near the base of panels, flexible or ductile connections allow for movement in the plane of the panel. These connections are usually threaded steel bars that have minimal resistance to movement perpendicular to their lengths, yet can withstand out-of-plane forces (Fig. 11.12). This kind of connection combination allows the structural frame to drift independently of the panel. Since the panels are not a full storey-height, the movement allowance need be less than for taller panels. Restraint against out-of-plane forces is shared by all four connections. The column cover-panels are attached to the columns using an identical approach except that the bearing connection is now near the bases of the panels. Horizontal gaps of between 10 mm to 20 mm between panels are usually sufficient to allow them to move freely relative to each other.

The diagrams are intended to illustrate the principles involved in the design of cladding panel connections. Other methods of connection are possible and will vary depending upon the components involved and the movement to be accommodated. In all cases appropriate architectural details need to be designed to addresses weather-tightness and other requirements of exterior movement joints.

Other materials

These claddings include stone slabs, ceramic tiles, insulated panels, thin sheet materials, such as fibre cement board and metal cladding in sheet and folded forms. For materials available in relatively small modules like stone slabs and ceramic tiles, drift is usually accommodated by anticipating a small horizontal movement along each horizontal joint. If joints are filled with a resilient sealant that can absorb movements without damage, special separation details might be unnecessary. Out-of-plane inertia forces per module are not difficult to resist. Where codes require that non-structural elements must not suffer damage during small earthquakes nor fall from a building during the design-level earthquake, then any proposed connection details need to be reviewed according to those two criteria.

Thin sheet cladding is widely used on commercial and residential buildings. Standard detailing usually allows for some minor movement, but specific detailing that incorporates separation joints between sheets may be required on buildings subject to large interstorey drifts (Fig. 11.13).

Irrespective of the cladding material used, the design of any cladding system in a seismically active region must cater for both interstorey drift and out-of-plane forces. The degree to which cladding elements are separated from the structure and each other depends largely upon the horizontal flexibility of the building. In some situations, particularly with lightweight cladding, wind effects might be more severe than those caused by earthquakes.

Windows and curtain-walls

Earthquake-induced interstorey drifts damage thin and brittle panes of glass. Glass breakage alone cost more than any other single item after the 1971 San Fernando, California, earthquake. More recently, during the 1994 Northridge, Los Angeles, earthquake up to 60 per cent of storefront glazing suffered damage in the worst affected areas. Little glazing damage was observed in high-rise buildings. From the dual perspective of injury prevention and reducing economic loss glazing is worth protecting.

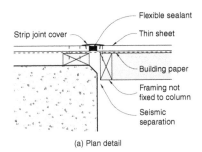

(a) Plan detail

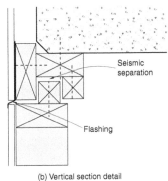

(b) Vertical section detail

▲**11.13** Separation details for thin sheet-cladding attached to a flexible structural frame.

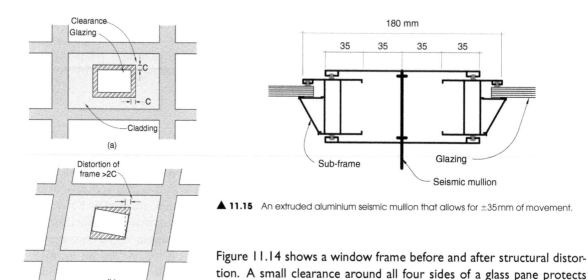

▲ **11.14** A glass windowpane set inside a frame with clearance all around (a). In a distorted frame the glass has slid and rotated relative to the base of the frame (b). Further distortion will lead to glazing damage.

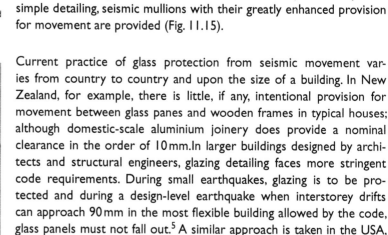

▲ **11.15** An extruded aluminium seismic mullion that allows for ±35 mm of movement.

Figure 11.14 shows a window frame before and after structural distortion. A small clearance around all four sides of a glass pane protects the glass under a small deformation imparted to the window frame during interstorey drift. Where drift exceeds that provided by such simple detailing, seismic mullions with their greatly enhanced provision for movement are provided (Fig. 11.15).

Current practice of glass protection from seismic movement varies from country to country and upon the size of a building. In New Zealand, for example, there is little, if any, intentional provision for movement between glass panes and wooden frames in typical houses; although domestic-scale aluminium joinery does provide a nominal clearance in the order of 10 mm. In larger buildings designed by architects and structural engineers, glazing detailing faces more stringent code requirements. During small earthquakes, glazing is to be protected and during a design-level earthquake when interstorey drifts can approach 90 mm in the most flexible building allowed by the code, glass panels must not fall out.[5] A similar approach is taken in the USA, where the drift at which a glass panel falls out is to be less than the calculated design earthquake drift multiplied by both a small factor of safety and one other factor reflecting the importance of the building.[6] Obviously provision for such large movements can pose practical and visual difficulties, particularly at corners where two-dimensional movements must be allowed for (Fig. 11.16).[7] During the design and specification of glazing, architects need advice from a structural engineer in order to provide the necessary amount of movement.

▲ **11.16** Damage to corner windows. Mexico City, 1985 Mexico earthquake. (Reproduced with permission from David C. Hopkins).

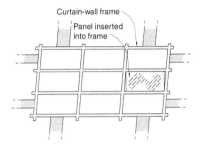

▲ 11.17   The elements of a curtain wall system.

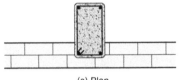

(a) Plan
Reinforced concrete column
provides face-load support

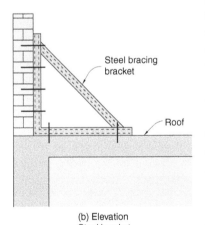

(b) Elevation
Steel bracket

▲ 11.18   Out-of-plane support to parapets using regularly spaced columns or brackets. Horizontal reinforcement in some mortar courses might be required to ensure the brickwork can span between vertical supports.

A lightweight cladding system consisting of glass, plastic or metal panels constrained within a light, often aluminium frame, is referred to as a curtain wall. The supporting frame is usually designed as a fully-framed prefabricated storey-height unit. It moves horizontally with the structural frame easily, offering little resistance (Fig. 11.17). Depending on the degree of interstorey drift seismic mullions may need to be incorporated into a curtain wall system. The basic approach is to isolate glass panes or other panels from their frames by providing suitably large movement clearances. Although simple in theory it is harder in practice to achieve the required separation. A high quality of workmanship is necessary to ensure that sliding or other movements are not hindered by tight-fitting gaskets or other devices. One set of tests found that the deflection capability of the tested panels was 40 per cent less than that calculated using a conventional formula.[8] Manufacturers' full-scale mock-up tests to demonstrate adequate seismic performance should be part of the process of design, specification and installation of curtain walls.

PARAPETS AND APPENDAGES

Parapets, particularly of unreinforced masonry construction, often fall from buildings during earthquakes. Due to high horizontal accelerations at roof levels parapets are vulnerable to out-of-plane collapse, especially if they cantilever vertically and depend upon unreliable mortar tension bond between block units.

Out-of-plane support can be provided by a variety of means. Vertical reinforced concrete cantilever columns or steel brackets can ensure the seismic safety of parapets if connected strongly to both parapet and roof structure (Fig. 11.18). There is usually no need to strengthen a parapet wall for inertia forces in its plane because of its inherent in-plane strength.

Appendages can take many forms including signs, canopies, cornices, mock-columns and other elements attached to the exteriors of buildings. In every case, appendages need to be tied-back or braced to structural elements that are strong and stiff enough to resist the inertia forces. The strength of a tie depends upon the intensity of the horizontal inertia force acting on an appendage, and that is related to the weight

▲ **11.19** Weakly attached unreinforced columns are tied back at first floor level with stainless steel ties and vertical bearing rods. Restaurant, Wellington.

▲ **11.20** Framing for interior partition walls. The wall top plate is inserted into a sheet steel channel without being nailed to it. This detail allows relative movement in the plane of the wall between the wall and the floor above. Office building, Wellington.

of the appendage, its height up a building and the dynamic characteristics of both itself and the building (Fig. 11.19).

PARTITION WALLS

Many of the comments made in relation to infill walls (Chapter 10) and masonry cladding above also apply to partition walls. If they are constructed from either reinforced or unreinforced masonry, they should be separated for in-plane movement yet restrained against out-of-plane forces. Given the risk of masonry partition walls toppling during earthquakes and endangering lives, out-of-plane restraint requires far more than a mortared joint between masonry units and structure (Fig. 10.13). A ductile detail using reinforcing rods is recommended (Figs 10.11 and 10.14). Alternatively, partition walls can be designed to be self-bracing by arranging for some panels to brace others provided they are all isolated from the interstorey drift of structure above them.

Light-weight partitions pose less danger to building occupants but still need to be separated to protect them from damage within flexible buildings (Fig. 11.20). Particular attention to detail is required to limit damage at the corners of partition walls.

Fire-rated partitions require special attention. They are unlikely to provide the necessary fire resistance if damaged by out-of-plane forces or interstorey drift. Given a feasible scenario of concurrent post-earthquake fire and fire sprinkler damage, the seismic detailing of these partitions must be of a very high standard and even more so when partitions are intended to provide safe egress from a building in the event of a fire.

SUSPENDED CEILINGS AND RAISED FLOORS

Injuries to building occupants, financial losses, and disruption of operations are all reduced by bracing suspended ceilings. Typically hanging from many lengths of fine wire, a suspended ceiling consists of a grillage of light metal members which support ceiling tiles. During an earthquake an unbraced ceiling swings about like a pendulum. It crashes against structure and other elements often damaging fire sprinkler heads and triggering deluges of water. Ceiling tiles dislodge

▲ 11.21 Suspended ceiling damage. 1994 Northridge, California earthquake.
(Reproduced with permission from A.B. King).

and fall, causing injuries and wrecking havoc on building interiors (Fig. 11.21).

Suspended ceilings, including the light-fixtures normally integrated within them, need bracing to prevent this uncontrolled swinging. Three techniques are illustrated in Fig. 11.22. The appropriate choice is based upon suspended ceiling manufacturers' technical information informed by full-scale testing and structural engineering calculations.

Figure 11.23 shows a method of bracing suspended ceilings to the floor soffit or roof structure above. This ensures good seismic performance. Contractors are reluctant to install bracing because of the extra effort required, so its necessity must be communicated clearly on architectural drawings, in specifications and followed through with site supervision to guarantee that it is correctly installed. The seismic integrity of a suspended ceiling depends largely on individual ceiling tiles providing in-plane diaphragm action (Chapter 4). The small clips that prevent tiles from uplifting and then falling are so easily omitted.

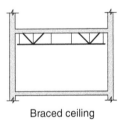

Braced ceiling

Lateral support from adjacent walls

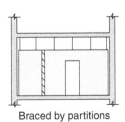

Braced by partitions

▲ 11.22 Three methods for bracing suspended ceilings to prevent them swinging during an earthquake.

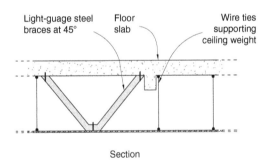

Light-guage steel braces at 45°   Floor slab   Wire ties supporting ceiling weight

Section

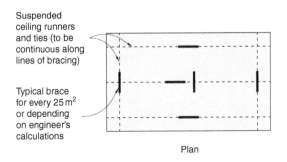

Suspended ceiling runners and ties (to be continuous along lines of bracing)

Typical brace for every 25 m² or depending on engineer's calculations

Plan

▲ 11.23 Detail and plan of a braced suspended ceiling.

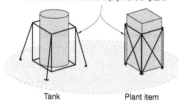

Protective bracing frame bolted directly to the floor around vulnerable equipment or plant

Tank   Plant item

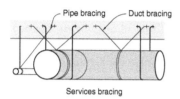

Pipe bracing   Duct bracing

Services bracing

▲ **11.24** Examples of mechanical plant restraints.

▲ **11.25** A braced roof-top water tank. Hospital, Whakatane, New Zealand.

These principles apply whether or not a ceiling is constructed from wood and plasterboard or is a proprietary modular system.

Raised floors that are installed in some office buildings provide space for, and allow access to, building services. Seismic restraint around the perimeter of the floors or cantilever action of their vertical supports provides resistance to horizontal accelerations. As for suspended ceilings, architects rely on a combination of manufacturers' information and structural engineering advice when specifying a proprietary raised floor system.

MECHANICAL AND ELECTRICAL EQUIPMENT

Mechanical equipment including water tanks, gas boilers, air conditioning equipment, pipes and other plant require restraint in order to prevent sliding and overturning during a quake. The aim is to prevent plant damage including rupturing of pipes (Fig. 11.24). Most buildings cannot continue to function when damaged equipment causes water damage. Since seismic accelerations near the roof of a multi-storey building can exceed 1.0g, serious damage is likely if plant is not bolted-down or braced to the main structure (Fig. 11.25). Plant damage avoidance is especially important for mechanical and electrical systems in critical facilities such as hospitals.

Seismic restraints should be provided for every item of mechanical and electrical plant. Structural engineers or mechanical engineers with seismic design experience will usually undertake the designs. This work may also include confirming or specifying that the item itself is robust enough to survive the intensity of anticipated shaking without internal damage or loss of function. Unfortunately, seismic restraints to equipment are frequently overlooked. A robust quality assurance system (as discussed in the introduction to this chapter) is one method to improve current practice.

BUILDING CONTENTS

Heavy and tall building contents slide and overturn when subjected to horizontal accelerations so these should be restrained; for example, high bookcases, storage cabinets and partial-height partition walls should be fastened to structural elements. These types of building

▲ **11.26** Damaged office contents. Mexico City, 1985 Mexico earthquake (Reproduced with permission from David C. Hopkins).

contents can cause considerable damage during an earthquake, and create significant risk of injury to occupants (Fig. 11.26).

As it is not feasible to restrain all building contents, attention should be focused upon those items most hazardous and important for post-earthquake function. A broad approach to both hazard and function is necessary. For example, a tall storage cabinet might be regarded as a hazard due its likelihood of overturning and causing injury. But it might also block egress by preventing a door from being opened, smashing bottles of hazardous chemicals stored on the floor, or damaging an adjacent computer file server. A rigorous program of restraining building contents, especially in a contents-rich building like one housing scientific laboratories, is challenging but necessary to minimize earthquake losses including loss of building function.[9]

Civil defence and emergency management agencies provide literature to help home owners and building occupants identify and restrain hazardous and valuable items. At least one code exists.[10] Specialist companies also offer seismic restraint devices and installation services to businesses wanting to minimize post-earthquake disruption or downtime and risk of injury to employees. Although most restraints are quite easily installed the real problem lies in their aesthetic and functional acceptance. Most of us are not used to seeing computers anchored to desks, or bench-top equipment tethered by chains or cables (Fig. 11.27). However, just as

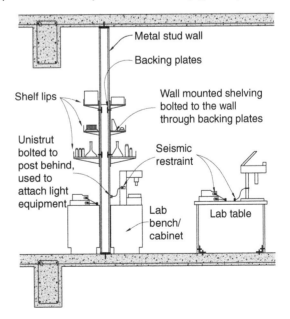

▲ **11.27** Restraints to laboratory equipment. (After Ref. 9).

seat belts are now worn in cars as a matter of course in many countries, the application of seismic restraints will become more common-place as the public's awareness of their importance grows.

REFERENCES

1 Lewis, J., Mohseni, A. and de Koning, M. (2006). Overcoming hurdles on the road to nonstructural seismic resilience for buildings. *Proceedings of the 8th US National Conference on Earthquake Engineering*, San Francisco. Paper 976, 10 pp.

2 Griffin, M.J. (2006). Earthquake performance of nonstructural components and systems: difficulties in achieving enhanced earthquake performance. *Proceedings of the 8th US National Conference on Earthquake Engineering*, San Francisco. Paper 1798, 10 pp.

3 Adan, S.M. and Tawresey, J.G. (2004). Retrofit of unreinforced masonry cladding following the 2001 Nisqually Earthquake. *Proceedings of the 13th World Conference on Earthquake Engineering*, Vancouver. Paper No. 1467, 12 pp.

4 Oliver, J. (2000). *The Brick Book*, Lifetime Books, Auckland.

5 Standards New Zealand (2004). *NZS 1170.5: 2004 Structural Design Actions, Part 5: Earthquake actions*. Standards New Zealand, Wellington.

6 Structural Engineering Institute/American Society of Civil Engineers (2005). *Minimum Design Loads for Building and Other Structures*, SEI/ASCE 7–05 including Supplement 1.

7 Massey, W. and Charleson, A.W. (2006). *Architectural Design for Earthquake: A guide to non-structural elements*. New Zealand Society for Earthquake Engineering, www.nzsee.org.nz.

8 Thurston, S.J. and King, A.B. (1992). *Two-directional Cyclic Racking of Corner Curtain Wall Glazing: Study Report No. 44*, Building Research Association of New Zealand.

9 Comerio, M.C. and Holmes, W.T. (2004). Seismic risk reduction of laboratory contents. *Proceedings of the 13th World Conference on Earthquake Engineering*, Vancouver. Paper No. 3389, 14 pp.

10 Standards New Zealand (1994). *Seismic Restraint of Building Contents*. Standards, New Zealand.

12 RETROFITTING

INTRODUCTION

The purpose of retrofitting is to reduce the vulnerability of a building's inhabitants and the building itself – its structure, non-structural elements and possibly its contents to earthquake damage. To retrofit a building is to improve its seismic performance. Alternative terms for retrofit, such as 'rehabilitation', 'upgrading', 'improvement', and perhaps 'strengthening', essentially convey the same meaning, but since 'retrofit' is well established, especially in the USA, it is used in this book. If the context of the term is unclear it can be preceded by 'seismic' to convey unambiguously one's involvement in one or more existing buildings possibly deficient from the perspective of seismic performance. The hesitation regarding the use of 'strengthening' is that sometimes it may be more appropriate to retrofit a building by merely increasing its ductility or even weakening selected structural members than by strengthening, but more on that later.

Retrofitting can be appropriate for any scale of building of any material. Wooden and stone houses are retrofitted, as are huge monumental concrete buildings. But, whereas wood-framed domestic construction, for example, can in many cases be retrofitted by their owners, larger projects are technically very challenging and require high levels of skill and experience in both the design and construction teams.

Although the retrofit or repair of earthquake-damaged buildings has been undertaken for centuries, the practice of retrofitting yet undamaged buildings in order to prepare them for a future damaging quake is far more recent. Thousands of buildings have already been retrofitted in California and New Zealand, as well as in other seismically active countries. Most retrofitting has yet to be put to the test by a large earthquake but evidence from at least one earthquake-damaged region confirms its value. 'Risk reduction efforts by the City of Los Angeles and neighboring communities greatly reduced the economic losses and threats to life in URM [unreinforced masonry buildings] during

▲ **12.1** Braced retrofit structure poorly integrated with the existing architecture, San Francisco.

▲ **12.2** A proposed perforated steel shear wall adds a layer of architectural richness to an interior wall of a historic unreinforced masonry building, Wellington.

the 1994 Northridge earthquake. In stark contrast, communities like Fillmore, which do not require URM risk reduction, suffered greater losses'.[1]

In contrast to the construction of a new building, retrofit begins with a process of assessment. Often undertaken in two stages, a preliminary assessment against pre-determined criteria is the basis for deciding whether any seismic deficiencies in a building warrant a more thorough investigation prior to recommending that retrofitting should or should not be undertaken. The subsequent steps of design and construction are also different from those of new construction due to their complexity. Designers and contractors usually find retrofit projects far more demanding than new construction. An existing building constrains their options and approaches. A successful retrofit scheme usually not only improves seismic performance but enhances functional and aesthetic building qualities.

Unfortunately much retrofit activity, at least in commercial and retail buildings, has failed to address aesthetic issues adequately. Look, Wong and Augustus, although writing in the context of retrofitting historic buildings summarize the situation:

'Although historic and other older buildings can be retrofitted to survive earthquakes, many retrofit practices damage or destroy the very features that make such buildings significant. Life-safety issues are foremost and, fortunately, there are various approaches which can save historic buildings both from the devastation caused by earthquakes and from the damage inflicted by well-intentioned but insensitive retrofit procedures. Building owners, managers, consultants, and communities need to be actively involved in preparing documents and readying irreplaceable historic resources from these damages'.[2]

Retrofits of all but historic buildings are usually driven by structural engineering and economic concerns rather than by architectural considerations. A cursory examination of retrofit schemes in downtown San Francisco reveals retrofit structure clashing with, and demeaning much of, the existing architecture (Fig. 12.1). Similar criticism, which can also be levelled at other cities, has initiated thinking that challenges the current approach often devoid of architectural merit.[3] Research-by-design retrofit proposals for a 1960s concrete frame office building and an early 1900s brick revivalist building give glimpses of alternative approaches indicative of a more architectural response (Figs 12.2 and 12.3).[4] While reading the following sections, which discuss retrofit issues pertaining to architects, keep in mind the desirability of retrofit solutions possessing architectural integrity.

▲ 12.3 A proposed retrofit scheme for a 1960s lift-slab commercial building, Wellington. The scheme references temporary propping to earthquake damaged buildings.

WHY RETROFIT?

Although a number of possible routes can result in a building being identified as requiring retrofit, the primary reason for retrofitting is to improve seismic performance. Retrofitting is recommended when a building is assessed as likely to perform poorly during moderate or greater intensity ground shaking. In most cases, poor performance will be caused by seismic deficiencies or the lack of one or more seismic-resistant features discussed in previous chapters. Possibly a building possesses insufficient strength, stiffness or ductility. Perhaps a discontinuous force path or a soft storey might lead to premature collapse. One or more of many potential reasons can lead to a building being retrofitted.

Identifying buildings requiring retrofit

But how is a building identified as a possible candidate for retrofit in the first place? The most obvious reason is that it has suffered damage during an earthquake. If the damage is severe enough to jeopardize building safety in a subsequent event then demolition or retrofit and repair are the only two options. However, the most common reason for retrofit is central or local government regulation. Many countries, states or cities require both new and certain existing buildings to comply with accepted seismic safety standards. For example, Los Angeles, San Francisco and Wellington have, for some years, taken pro-active steps to improve their communities' seismic resilience by reducing vulnerability to seismic hazards. San Francisco has published a brochure for building owners of unreinforced masonry buildings advising them of the steps to be taken towards retrofit.[5] Retrofitting is also often required when an under-strength building undergoes a change of use. Increasingly, state or city laws require old buildings and even those more recent yet seismically vulnerable, but usually excluding houses, to meet certain minimum standards.

Old buildings are targeted for several obvious reasons. In the light of today's knowledge past codes of practice to which buildings were originally designed are out-dated. The concept of ductility, that structural characteristic most likely to ensure earthquake survivability, was tentatively first introduced into codes around the 1960s. Capacity Design was first codified in the mid-1970s. Until then, designers lacked knowledge and guidelines on how to prevent a building from collapsing when seismic forces exceeded its design strength. As in all professions, new knowledge is continually introduced, leading to improved practices.

▲ 12.4   A damaged reinforced concrete building, 1995 Kobe, Japan earthquake. (Reproduced with permission from Adam Crewe).

Another reason for focusing attention on older buildings is because some types have performed poorly in recent earthquakes. The 1995 Kobe earthquake highlighted, perhaps more dramatically than the preceding Californian 1989 Loma Prieta and 1994 Northridge earthquakes, the vulnerability of pre-mid 1970s non-ductile reinforced concrete buildings (Fig. 12.4). The Northridge earthquake was notorious for exposing the deficiencies of steel moment frame buildings. These lessons, along with others learnt from research computer simulations and laboratory test programs, have led to improvements in current codes. Design professionals and also politicians have been alerted to the seismic vulnerability of a significant portion of their communities' building stock.

Other reasons to retrofit

A climate of increasing public expectation for building safety has also led to higher mandatory standards. In New Zealand all deficient school buildings have been identified and retrofitted. A similar program is underway in British Columbia, Canada. A California Senate Bill 'mandates that all health care facilities providing acute services be retrofitted to life-safety performance level by 2008, and a full serviceability level [fully functional after the design-level earthquake] by 2030'.[6]

Not all retrofits though are driven by legal requirements. Some building owners take the initiative themselves. A decision to retrofit reflects the real possibility of business failure or severe financial losses after a damaging quake critically interrupting commercial operations. Other owners, like museum boards, desire to protect their collections. An increasing number of Californian homeowners retrofit their houses as an alternative to purchasing what they consider to be very expensive earthquake insurance.

Assessment

As mentioned previously, assessment is the first step in the process of retrofitting. The seismic vulnerability of a building must first be ascertained before deciding whether or not to retrofit. Building assessment usually consists of two stages. An initial or preliminary assessment begins the process. Some countries have developed assessment procedures based upon extensive knowledge of their local building types, construction methods and materials and history of seismic code developments.[7,8] After a brief visual inspection to identify any serious structural weaknesses, such as critical configuration problems, structural engineers 'score' a building, expressing its strength as a percentage

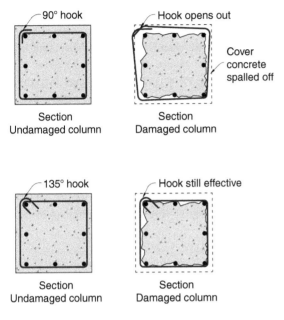

90° hook

Hook opens out

Cover concrete spalled off

Section
Undamaged column

Section
Damaged column

135° hook

Hook still effective

Section
Undamaged column

Section
Damaged column

▲ **12.5** Incorrectly bent column (and beam) ties are dangerous. During an earthquake their hooks open out and the ties lose all their effectiveness.

of current seismic code requirements. Depending on the result, a decision is made either not to retrofit or to proceed to a second level of assessment which involves a far more thorough procedure. An on-site inspection, possibly involving some minor demolition to determine welding or reinforcing steel details, is followed by extensive structural engineering analysis. Only then is a final decision made on whether or not to retrofit.

Poor structural configuration, such as short-columns or a soft-storey, often necessitates retrofit. Inadequate detailing can be another reason. Poor details, like incorrectly bent column ties in a reinforced concrete frame building, effectively rob their columns of most of their shear strength and confinement during a design-level earthquake. Such a small detail – 90 verses 135 degree tie hooks – has such severe consequences for a building's seismic safety (Fig. 12.5). Inferior construction materials, such as under-strength concrete or reinforcing steel that is too weak or brittle, can also be an issue. The quality of the construction might be poor. Column or beam ties might be missing, reinforcing bars not lapped properly or placed incorrectly – the list of potential problems is lengthy. Some problems on their own may not be too serious. However, others definitely require remedying.

RETROFIT OBJECTIVES

Once a building has been assessed as requiring retrofit what are the retrofit objectives and standards? In the case of voluntary owner-initiated retrofits, the objectives can be decided by the owner in conjunction with his or her design team. However, in mandatory programs, objectives are prescribed by the relevant public agencies. Objectives are not uniform for all buildings but are related to building function. Hospitals and fire stations, for instance, must satisfy far more rigorous performance standards than ordinary, say, office buildings.

Ideally, 'the primary goal [of a retrofit].... is to provide a level of safety for rehabilitated buildings similar to that of buildings recently designed to US seismic code requirements'.[9] New Zealand guidelines contain comparable expectations, recommending that a retrofit should bring the building to 'as near as is reasonably practicable' the standards pertaining to a new building.[8] These guidelines note that, although the legal threshold requiring retrofit is under 34 per cent of current code standards, any improvement should be at least to 67 per cent of current code. The effect of bringing a building from 33 to 67 per cent of current code is to reduce its risk of severe damage from 25 times to 5 times that of a new building. The Wellington City Council lists the following benefits to encourage building owners to adopt high retrofit standards:

- improved levels of safety for occupants, tenants and the public
- allowance for a change in use to occur to potentially better meet owner or market demand and realize a better return
- insurance against future changes in either the legislation or structural codes which may require higher levels of strengthening to be achieved
- leverage for improved insurance
- reduced risk level of damage to the building, other properties in its proximity and lessen the impacts on business continuity.[10]

Retrofit objectives, while led by structural engineering imperatives, should also include reference to architectural issues. While not mandatory, the architectural implications of various retrofit alternatives need to be discussed by the building owner, architect and engineers. How do the retrofit alternatives impact on the façade and building interior? Should any new retrofit structure be expressed? If so, what structural language, what materials and detailing? These and other questions need to be addressed together with associated financial implications.

RETROFIT APPROACHES

Approach the retrofitting of a building with the aim of minimum structural and architectural intervention. Only after all of the existing structural strength and ductility has been mobilized and still found lacking, should designers consider a more invasive approach; like the provision of new seismic resisting systems.

Improving seismic performance

Minimum intervention involves utilizing the capabilities of the existing structure to the greatest extent possible. This might be achieved by improving its performance sufficiently by strengthening one or two individual members or connections rather than inserting a whole new structural system. But first, the adequacy of the strength, stiffness and ductility of all structural elements and connections in seismic

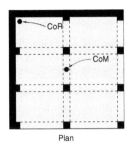

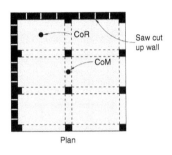

▲ 12.6 By weakening the structural walls, the centre of rigidity (CoR) is brought closer to the centre of mass (CoM), reducing torsional eccentricity.

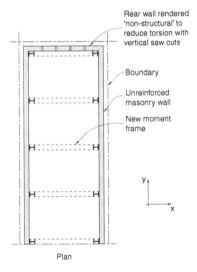

▲ 12.7 Y direction seismic forces are resisted by long masonry walls but new moment frames are required to resist x direction forces. Frame columns also provide out-of-plane resistance to the masonry walls.

force paths require evaluation. Remember, a continuous force path is required for seismic forces acting in the direction of each of the two main orthogonal axes of a building (Chapter 2). Below is a typical list of force path components a structural engineer checks in detail in each direction:

- Strength of exterior and interior walls against out-of-plane forces (Chapter 2)
- Connections of those walls to diaphragms
- Diaphragms (Chapter 4)
- Connection of diaphragms to primary vertical structure like shear walls
- Primary vertical structure, such as shear walls, braced frames or moment frames (Chapter 5)
- Connection of vertical structure to the foundations, and
- Foundations (Chapter 7).

At this stage of structural review, the existing building is also checked for any horizontal or vertical configuration problems that might compromise its seismic performance (Chapters 8 and 9). Irregularities, such as diaphragm discontinuities in the form of large openings, significant torsional eccentricity or a soft-storey, may be too severe to deal with by merely improving what exists.

If the results of force path and configuration evaluations indicate that a strategy of improvement rather than renewal will be successful, then those members and connections found deficient are upgraded using one or more of the reasonably standard techniques outlined in the following section. Occasionally, some members may even warrant intentional weakening. For example, the torsional response of a building with very strong eccentric boundary walls might be improved by vertical saw cuts to 'soften' them up (Figs 12.6 and 12.7).

The approach of improving an existing structure is particularly appropriate for seismically sub-standard housing. Relatively minor upgrading, like bolting walls down to the foundation or increasing the bracing in the foundation crawl space by strengthening perimeter or cripple walls, is effective. Many publications by public bodies aimed at homeowners are available on the internet.[11] At least one US city provides earthquake strengthening workshops, handbooks for homeowners, free house retrofit plan sets, a list of contactors who completed the city's home-retrofit Contractor Workshop, a construction tool lending library and limited financial assistance.[12]

Provision of new structural systems

Where an existing structure is clearly inadequate, one or more new seismic resisting systems may need to be inserted. Usually one of the three primary vertical structural systems discussed in Chapter 5 is chosen for each orthogonal direction, mindful of the need for structural compatibility. The new system must be stiff enough to resist seismic forces before its horizontal drift damages the existing structure and fabric excessively. For this reason, relatively flexible moment frames rarely strengthen seismically inadequate shear wall buildings. Another structural compatibility concern involves the ability of the existing gravity force-bearing structure to drift the same amount as the new seismic structure without sustaining serious damage. Gravity structure, by definition, is not expected to resist seismic forces. Nonetheless, it must be capable of enduring the drifts imposed upon it as the new structure flexes in a quake.

For an example of where just one new seismic system is required, consider long and narrow buildings fronting onto a street. Often their concrete or even unreinforced masonry boundary walls may be sufficiently strong to resist longitudinal (y direction) forces though the out-of-plane resistance of those walls may need improvement. However, a completely new structural system may be necessary in the direction parallel to the street (Fig. 12.7).

Sometimes it is enough to modify an existing structure rather than provide new structural systems. The insertion of new structural elements into an existing building can transform one structural system into another. Take the case of deficient reinforced concrete moment frames. It is possible to insert diagonal steel braces to form composite concrete and steel braced frames with potentially enhanced strength and ductility. Alternatively, a moment frame can be infilled with reinforced concrete to form a concrete shear wall (Fig. 12.8). Where such transformative approaches are taken, inserted or infill elements must rise from the foundations continuously up the height of a building to avoid a soft-storey. Foundation upgrades can be expected due to the increased forces these stronger and stiffer modified structures attract.

Weight reduction

Irrespective of the degree of retrofit intervention, architects and engineers should always try to reduce building weight. As discussed in Chapter 2, the less weight the less seismic force. Just as health professionals encourage us to reduce excessive body weight, so too with

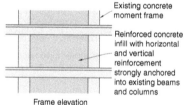

Existing concrete moment frame

Reinforced concrete infill with horizontal and vertical reinforcement strongly anchored into existing beams and columns

Frame elevation

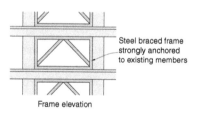

Steel braced frame strongly anchored to existing members

Frame elevation

▲ **12.8** Transformation of a concrete moment frame into a shear wall and braced frame.

buildings. The removal of heavy elements like masonry partitions, parapets and chimneys can reduce retrofit intervention and cost, provided that desirable architectural qualities are not sacrificed.

Adjacent buildings

Separation gaps provided between new construction and boundary lines (as discussed in Chapter 8) are frequently not present between existing buildings. If any gaps have been provided, their widths are likely to be far less than those required by current codes. Although the likelihood and severity of pounding between adjacent buildings can be reduced by structural stiffening during retrofit to reduce seismic drift the problem is unlikely to be eliminated. Since widening an existing gap between buildings is usually impractical, designers are left with the worrying possibility of pounding. Pounding can seriously damage perimeter force-bearing elements like columns and walls, especially if the floor slabs of adjacent buildings do not align. Designers should at least provide alternative force paths, such as supplementary columns or props located away from the potential damage zone so that in the event of pounding damage gravity loads continue to be safely supported.

Retrofit techniques

An extensive range of options is available to architects and structural engineers who retrofit a building by improving its existing structure. Some of the more common techniques applicable to elements like columns and beams are illustrated in Fig. 12.9. Where new horizontal and vertical structural systems are required they are designed and detailed as outlined in Chapters 4 and 5. They are essentially identical to systems utilized in new buildings, but need to be structurally connected to the existing construction. Often they require new or upgraded foundations to prevent overturning. The more slender the vertical elements, like shear walls or braced frames, the greater the vertical axial forces to be resisted at foundation level. New systems – as well as seismically improved existing systems – must be configured in-plan to not only resist seismic forces in two orthogonal directions but also to resist torsion, as discussed in Chapters 2 and 8.

Unreinforced masonry walls

Whether functioning as load-bearing infill walls or free-standing partition walls these heavy and brittle walls are vulnerable to out-of-plane forces. Many of these wall types collapse in earthquakes. They require

Description	Method	Comment
Concrete column: - steel jackets and straps	Existing column (reinforcement not shown) Grout or concrete Steel encasement Section Continuous steel angle Steel strap Section	A steel jacket encases column confinement concrete, and increases shear strength and ductility. Steel straps can be passed through holes in beams to confine and strengthen beam-column joints.
Concrete column: - composite fibre wrapping	Chamfered corner Carbon or glass fibre bedded in an epoxy or other bonding material Section	Composite fibre wrapping increases column confinement, shear strength and ductility.
Concrete column: - concrete jacketing	Vertical bar Continuous (welded) horizontal tie Cross-tie Concrete Section	Concrete jacket increases column confinement, bending and shear strength, and ductility. Vertical reinforcement must be anchored into the foundation and continuous through floor levels. Beam-column joints are strengthened by passing cross-ties through existing beams.
Concrete column: - strengthened with side walls	Existing column New side wall Section	Side walls, if continuous up a column, can modify a dangerous weak column—strong beam moment frame, into a strong column—weak beam frame.
Concrete beam: steel hoops	Hole cut through the slab Existing beam Steel strap (fully welded) Section	Welded steel straps at close enough centres confine the beam and increases its shear strength. If the beam corners are chamfered it is possible to use composite fibre hoop wraps.

▲ **12.9** Methods of retrofitting concrete columns and beams.

Description	Method	Comment
Post/mullion support to unreinforced walls	Unreinforced masonry wall Steel post/mullion, UC or UB Bolt grouted into wall *Vertical section*	Posts require strong connections to upper and lower diaphragms. This method provides walls with out-of-plane resistance.
Concrete skin on concrete or unreinforced masonry shear walls	Existing wall Grouted-in tie Reinforcing bar Sprayed-on (shotcrete) or cast-in-place concrete *Vertical section*	A concrete skin increases the strength of an existing shear wall and provides out-of-plane resistance
Composite fibre or steel mesh overlay to unreinforced concrete or masonry shear walls	Fibre or steel mesh Plaster bedding *Vertical section*	A double-skin overlay increases in-plane shear wall strength as well as out-of-plane resistance. Fine steel mesh in cement plaster is also effective. Individual fibre-strips can be used also.
Post-tensioned unreinforced shear walls	Post-tensioned tendons *Vertical section*	Post-tensioning requires vertical holes drilled down walls into the foundation where tendons are well anchored. Post-tensioning increases shear wall strength and out-of-plane resistance. An alternative detail is to position tendons outside a wall on either side.

▲ **12.10** Methods of retrofitting unreinforced concrete or masonry walls.

support. Vertical structural members, usually steel posts or mullions, are inserted into the walls or fixed to one of their faces (Fig. 12.10). Frequent and strong connections from these members into the masonry are necessary. The posts must also be strongly connected top and bottom to diaphragms, which then transfer horizontal forces from the posts to principal seismic resisting systems elsewhere in plan of the building. New reinforced concrete columns can also provide out-of-plane resistance to unreinforced masonry walls. In a multi-storey building place these vertical elements, typically installed between 3 m and 5 m apart, above each other. Then, if an area of masonry is severely damaged, this new vertical structure can also act as props, preventing collapse of the floors of the building in the vicinity of the wall damage. This new vertical structure then functions both as a mullion as well as a gravity prop.

Another method to provide out-of-plane resistance is to apply a thin coat of plaster reinforced by a steel or fibre mesh to each side of a wall. This creates a sandwich panel capable of spanning vertically between floor diaphragms. If a coating can be plastered on one side only, a thicker layer of up to 200 mm reinforced concrete, cast-in-place or sprayed as shotcrete can provide adequate strength (Fig. 12.11). Both types of layered strengthening also contribute to the in-plane strength of a retrofitted wall.

▲ **12.11** Shotcreting to form a new internal reinforced concrete wall, Wellington.

Diaphragms

Existing diaphragms often require upgrading. Particularly in an unreinforced masonry building with wooden floors, it is not feasible to structurally improve existing low-strength diaphragms. In these cases, new diaphragms are constructed above or below existing floors or ceilings. One method that involves casting a new reinforced concrete slab over the existing flooring, provided that the floor joists can support the extra weight, adds undesirable additional weight to the building. A lighter alternative diaphragm is fabricated from structural steel to form a braced diaphragm or horizontal truss. Where a diaphragm resists and distributes a lesser amount of seismic force from surrounding walls to vertical structural elements a new plywood diaphragm can be laid over or under existing flooring or roof framing and diaphragm collector and tie members upgraded as necessary. Fig. 12.12 shows typical diaphragm retrofit solutions.

Shear walls

Of all primary vertical systems, reinforced concrete shear walls provide the best retrofitting option. They have, as mentioned in Chapter 5,

Description	Method	Comment
Timber floor diaphragm overlay	New plywood sheet · Existing flooring · Joist	An overlay with plywood sheeting increases diaphragm stiffness and strength. Sheet-to-sheet and sheet-to-existing flooring connections require careful attention.
Braced Steel strap diaphragm	New flooring packed above straps · Steel strap · Joist	Diagonal steel straps in conjunction with existing floor members form a horizontal truss (diaphragm). The straps can also be placed under the joists.
Diaphragm to wall connections. Use either A or B	A · B · Steel angle bolted regularly to wall · Plywood sheet overlay · Existing floor · Existing joist · Steel strap bracing · Steel channel — Vertical section	Where the steel angle or channel is continuous it also functions as a diaphragm chord. Detail A is used when diaphragm strengthening is confined to above the floor, and B, below.
Concrete overlay diaphragm	New concrete overlay · Reinforcing mesh — Section	A reinforced overlay increases the strength and stiffness of an existing diaphragm.

▲ **12.12** Methods of retrofitting diaphragms.

by far the best seismic track record of the three systems. Construction is easier if walls are placed on the outside of a building where their new foundations (which are often required) are much easier to construct. If exterior walls are chosen, pay careful attention to their architectural impact on the existing building. For buildings with configuration deficiencies (like soft storeys) it may be sufficient to provide a 'strong back' wall. This is a conventional shear wall, except that it can be pinned at its base or allowed to rock, thus greatly reducing foundation costs.[13]

▲ **12.13** External retrofitted concrete shear walls, Vancouver.

Figure 12.13 shows an example of a building retrofitted by external shear walls. Two pairs of reinforced concrete shear walls now resist all seismic forces parallel to their lengths. They free the seismically deficient 1967 riveted steel moment frames from needing to resist seismic forces. The walls are strongly attached to each floor diaphragm and at their base are cast into a deep basement beam that reduces their tendency to overturn under seismic forces. Soil anchors at each end of the beam prevent it overturning.[14]

In a second example (Fig. 5.8), an increasing number of windows up the height of the wall reflect the reduction in shear force. Large and deeply embedded tension piles under the wall prevent it from overturning and help transfer horizontal seismic forces into the soil.

Braced frames

Braced frames, usually steel, are among the most common and cost-effective primary retrofit systems. Compared to reinforced concrete walls they are light-weight and have less impact upon views from windows and natural light. Figure 12.14 shows rather unusually configured, yet fully triangulated braced frames. Most load transfer is through tension forces in the slender diagonal members as their buckling capacity is limited. Eccentrically braced frames inserted into a 1960s three-storey reinforced concrete building provide a new line of seismic resisting structure (Fig. 12.15). Strong

▲ **12.14** External retrofitted braced frames, Vancouver.

▲ **12.15** Eccentrically braced frames strengthen a reinforced concrete building, Wellington.

connections at diaphragms transfer forces into the new frames. Braced frames incorporating buckling-restrained braces are increasingly an alternative to eccentric braced frames (Chapter 14).

Moment frames

A 1950s reinforced concrete frame building – strengthened by the addition of new steel moment frames – is illustrated in Fig. 12.16. Located at the rear of the building, the new frame connects to the existing construction by grouted-in bolts at every floor. On-site bolted joints at the mid-span of beams where no bending moments occur during an earthquake are visible. A similar frame is inserted inside the front façade at the other end of the building. New reinforced concrete frames provide lengthwise stability for the building shown in Fig. 12.17. During the retrofit, the existing exterior columns were transformed by the layering of vertical slabs to form new moment frame columns.[15]

New beams that also function as balustrades at every alternate storey form a moment mega-frame. Although prior to the retrofit every balcony slab had a transparent metal balustrade this scheme is a good example of an architecturally well-integrated retrofit. On the same campus another moment frame retrofit also respects, if not improves upon, the existing architecture (Fig. 12.18).

▲ **12.16** An external steel moment frame retrofit, Wellington.

▲ **12.17** New exterior columns and beams form a double-height moment mega-frame, Berkeley, California.

▲ **12.18** A retrofit solution consisting of exterior moment frames. The surface markings identify remedial work to surface coatings, Berkeley, California.

Seismic isolation

Seismic or base-isolation is another retrofit technique that involves the insertion of new structural elements. Of all retrofit schemes it probably involves the most severe intervention. But it offers the best seismic performance in terms of mitigation of seismic damage to both the fabric of a building and its contents – hence its popularity when retrofitting historic buildings, museums, and galleries.

In accordance with the principles discussed in Chapter 14, a base-isolation retrofit involves severing the entire building from its foundations and inserting horizontally flexible bearings under gravity load-bearing columns and walls. Considerable foundation upgrade work is often required, as well as new beams just above the bearings. These beams are needed to transfer gravity forces from the walls into the bearings which are placed approximately 5 m apart. Unfortunately, even with an isolation system in place, the existing superstructure often requires additional strength. New reinforced concrete shear walls or steel braced frames are typically inserted above the isolation plane to help the existing structure resist inertia forces and transfer them down to the isolators and foundations. Several examples of base-isolation retrofits are shown in Figs 14.11 to 14.14.

Non-structural retrofit

As explained in the previous two chapters, many buildings contain non-structural elements that are potentially hazardous to people during an earthquake and expensive to repair. Some elements – like infill walls – are even more dangerous as they can cause severe structural damage, endangering a whole building. In order to meet the minimum life-safety or higher performance goals of a retrofit, non-structural elements require attention.

Masonry infill walls

As well as requiring strengthening against out-of-plane forces, infill walls may also need special attention due to their ability to function, even though completely unintended, as structural elements. The structural engineer and architect, as discussed in Chapter 10, are faced with two choices. Either accept and utilize infills as structural members, or physically separate them from their frames to prevent them functioning structurally while protecting them against out-of-plane forces.

In any structure, partial-height infills are dangerous because they shorten the effective length of a column causing a brittle short-column effect (Chapter 9). These infills should be separated by making vertical cuts close to the columns they infill after the structural engineer has calculated the width of the separation gaps. Gaps must be treated architecturally to achieve satisfactory weathering, acoustic and fire performance. If a new shear wall resists seismic forces then due to its stiffness infills may not require attention related to their interaction with surrounding frames.

Staircases

Since the rigidity of staircases can cause damage to the structure as well as to themselves, they may also require separation from the main structure (Chapter 10). Retrofit details should allow inter-storey drift to occur between floors in any direction without restraint from staircases.

All other functionally important or hazardous elements

All the measures suggested in Chapter 11 are relevant to retrofitting such elements. FEMA-172 contains additional details.[16]

HISTORIC BUILDINGS

The retrofit of historic buildings invariably requires a variety of conservation approaches. Any retrofit scheme must be consistent with, and fully integrated with, the chosen approach. For example, if the form and materials of an existing building are to be preserved, retrofitting techniques might need to be concealed. This may require the use of more innovative and sophisticated retrofit methods than normal. On the other hand, full or partial exposure of retrofit systems and details may be acceptable where a building's rehabilitation involves more general preservation of architectural, cultural and historical values, possibly including former alterations and additions. Total concealment of retrofit structure may or may not be required when restoring a building to its original condition.

Whatever one's attitude towards conservation, far greater architectural sensitivity to the retrofit process and outcome is needed. Technical strengthening requirements need to be balanced with principles of architectural conservation. Look, Wong and Augustus recommend adopting the following three principles:

- Historic materials should be preserved and retained to the greatest extent possible and not replaced wholesale in the process of seismic strengthening;

- New seismic retrofit systems, whether hidden or exposed, should respect the character and integrity of the historic building and be visually compatible with it in design; and
- Seismic work should be "reversible" to the greatest extent possible to allow removal for future use of improved systems and traditional repair of remaining historic materials.[17]

The three authors do not insist that retrofit structure be concealed, nor recommend as some do, that strengthening elements be located in spaces of least historic value. After all, if retrofit structure is well-designed and respectful of the existing fabric, it can add another layer of historical intervention and avoid perpetuating the myth that the original construction is earthquake-resistant. This unconventional approach, which was investigated in a design project where exposed retrofit structure enriches the interior architecture of a historic unreinforced masonry building, was discussed in the introduction to this chapter (Fig. 12.3).[4]

Whereas the main seismic performance goal for most buildings requiring retrofit is to preserve life and minimize injuries to occupants, the conservation plan prepared prior to an historic building retrofit may require designers to consider a second goal – to protect the building fabric. Although each historic building has to be treated individually, there may be a desire to control or limit the amount of damage suffered in the design-level earthquake. It might be inappropriate to adopt a lower performance criterion of life safety, where, although no lives are lost during a quake, a building is so badly damaged it requires demolition. However, as Randolph Langenbach points out, if seismic performance standards are too high, ironically the ensuing retrofit interventions may compromise or even destroy the very fabric to be protected. He argues for an approach to seismic retrofit that is far more sensitive to preservation ideals than much current practice:

> *Unlike maintenance and rehabilitation from decay, a seismic project may tear apart a building which was otherwise in good repair and make it almost entirely new. In such an instance, only the image, rather than the substance, of much of the historic fabric is preserved.*[18]

REFERENCES

1 Seismic Safety Commission (1995). *Status of the Unreinforced Masonry Building Law: 1995 Annual Report to the Legislature.* Seismic Safety Commission, California.

2 Look, D.W., Wong, T. and Augustus, S.R. (1997). *The Seismic Retrofit of Historic Buildings: Keeping preservation in the forefront, Preservation Brief 41.* National Park Service USA. Available at www.cr.nps.gov/hps/tps/briefs/brief41.htm.

3   Charleson, A.W. and Taylor, M. (1997). Architectural implications of seismic strengthening schemes. In *Structural Studies Repairs and Maintenance of Historic Buildings*, Sanchez-Beitia, S. and Brebbia, C.A. (eds.). Computational Mechanics Publications, pp. 477–786.

4   Taylor, M., Preston, J., and Charleson, A.W. (2002). *Moments of Resistance*. Archadia Press, Sydney.

5   SF DBI (2001). *What you should know about unreinforced masonry buildings*. San Francisco Department of Building Inspection.

6   Bruneau. M. et al. (2005). Review of selected research on US seismic design and retrofit strategies for steel structures. *Progressive Structural Engineering Materials*, No. 7, 103–114.

7   Applied Technology Council (1988). *Rapid visual screening of buildings for potential seismic hazards: a handbook. ATC–21*. Federal Emergency Management Agency.

8   NZSEE Study Group on Earthquake Risk Buildings (2006). *Assessment and improvement of the structural performance of buildings in earthquakes: prioritisation, initial evaluation, detailed assessment and improvement measures*. New Zealand Society for Earthquake Engineering.

9   FEMA (1997). *NEHRP guidelines for the seismic rehabilitation of buildings: FEMA-273*. Federal Emergency Management Agency.

10 WCC (2005). *Draft Earthquake-prone Buildings Policy*. Wellington City Council.

11 For example, CSSC (2005). *Homeowner's Guide to Earthquake Safety*. California Seismic Safety Commission.

12 San Leandro City (2005). *Earthquake Retrofit Programs*. http://www.ci.san-leandro.ca.us/cdearthretro.asp.

13 NZSEE Study Group on Earthquake Risk Buildings (2006). *Assessment and improvement of the structural performance of buildings in earthquakes: prioritisation, initial evaluation, detailed assessment and improvement measures*. New Zealand Society for Earthquake Engineering, pp. 13–16.

14 Sherstobitoff, J., Rezai, M. and Wong, M. (2004). Seismic upgrade of Lions Gate Hospital's Acute Tower South. *Proceedings 13th World Conference on Earthquake Engineering*, Paper No. 1423. 12 pp.

15 Comerio, M.C., Tobriner, S. and Fehrenkamp, A. (2006). *Bracing Berkeley: A guide to seismic safety on the UC Berkeley Campus*. Pacific Earthquake Engineering Research Centre, University of California.

16 FEMA (1992). *NEHRP handbook of techniques for the seismic rehabilitation of buildings: FEMA-172*. Federal Emergency Management Agency.

17 Look, D.W., Wong, T. and Augustus, S.R. (1997). *The seismic retrofit of historic buildings: keeping preservation in the forefront, Preservation Brief 41*. National Park Service USA. Available at www.cr.nps.gov/hps/tps/briefs/brief41.htm.

18 Langenbach, R. (1994). Architectural issues in the seismic rehabilitation of masonry buildings. *Proceedings of the US-Italy Workshop on Guidelines for Seismic Evaluation and Rehabilitation of Unreinforced Masonry Buildings*; University of Pavia, Italy, June 22–24. Also in NCEER Report 94-0021. National Center for Earthquake Engineering Research, State University of New York at Buffalo, 1994, 1–3 to 1–16.

13 PROFESSIONAL COLLABORATION AND COMMUNICATION

INTRODUCTION

It is most unusual for an architectural book on a structural theme to devote an entire chapter to the subject of collaboration and communication. While these aspects are important for the successful design and construction of any building, they become especially critical in earthquake-prone regions where seismic forces so often dominate structural considerations. In seismically benign regions, architects rarely need to discuss structural issues with a client, or to have such a strong collaborative approach with others in the design team. Special 'structural' discussions with contractors are also held infrequently. But seismic resistant design and construction is different in many ways from non-seismic approaches to the structuring of architecture. It is much more difficult to outwit a quake than resist gravity or even wind forces.

So what sets seismic design apart? First, a client will probably not appreciate that a new building, even though complying with all applicable codes and regulations, can certainly not be considered 'earthquake proof' or be totally resistant to seismic damage. In fact, if a building possesses the minimum possible structure for seismic resistance it is designed for ductility, and that means damage. Secondly, as an architect you may be disappointed, if not shocked, at how much seismic resisting structure your project requires. Its structural footprint may far exceed what you have come to expect from studying building structures in overseas journals. And finally, you have to deal with contractors' scepticism or downright reluctance to follow details on the plans that seem almost counter-intuitive and unnecessarily complicated – such as separation gaps between infill walls and frames, sacrificial non-structural detailing across seismic joints and concentrations of reinforcing ties in beam-column joints and at potential structural fuse regions. The successful resolution of all the above issues and many more require intentional and sometimes intense collaboration and communication.

An architect, for all but the smallest and most simple buildings when he or she may work alone, is a member of a design team. Figure 6.5 proposes a model whereby an architect considers structural and, therefore, seismic issues at a conceptual design stage. Only such early engineering input achieves an outcome where planning, structure and services are integrated and support the architectural concept and program. The success of this approach depends on mutual respect and cooperation between team members. In many cases, as project leader, an architect must both demonstrate and encourage clear communication and common understanding between team members and try to harness their personalities and skills for the sake of the project. Similar positive teamwork attributes also need to be carried through into the construction team, formed when the contractor and design team join forces.

CLIENT

As mentioned above, conversations between client and architect rarely include structural issues. Perhaps the need for some floor areas to support heavy file-storage systems warrants discussion, but otherwise a client expects the architect and structural engineer to deliver an economic and trouble-free structure. After all, the client is familiar with the concept of gravity forces. A gravity resisting structure is experienced every day of the week although mostly at a subconscious level. And, unless accommodated in a building noticeably flexible during wind gusts, the client may be disinterested in details regarding wind design.

In total contrast, for the sake of both client and architect, seismic design issues need explicit and collaborative attention. It is not enough for an architect to merely satisfy the legal obligations set by city authorities and comply with standard building code requirements. Clients need to be informed regarding the basis of a seismic design and the expected seismic performance of the building. Any unrealistic client expectations, such as the building remaining undamaged in high intensity shaking, need to be discussed and probably dispelled.

Ideally, at the time an architect presents a concept sketch to a client the seismic design assumptions for the building are outlined. Include the structural engineer in this presentation since some deep issues are likely to be raised. In this situation the architect plays an educative role, explaining to the client some of the features of seismic resistant design discussed in Chapter 3. Most clients are surprised to hear that their new ductile building may be designed for as little as 15 per cent of the actual seismic forces expected during a design-level earthquake even though it complies with current structural codes. Advise them of the probability of the design-level earthquake occurring during the life of the building and the extent of damage likely to be sustained. This discussion will also raise questions like: 'Will the damage be repairable and at what cost?' and 'How long might the building be out of operation following a quake?' The seismic performance of non-structural components also needs exploration at this stage of the design, at least in a general way.

Table 13.1 facilitates discussion of seismic standards and building performance.[1] Before meeting a client, the structural engineer should shade

▼ **13.1** A checklist to facilitate discussion of earthquake expectations between architect and client (Adapted from FEMA 389, 2004)

EARTHQUAKE EXPECTATIONS OF CLIENT

A. Earthquake performance of structure

Intensity of shaking	Damage			
	Severe to non-repairable	Repairable: evacuation	Repairable: no evacuation	No significant damage
Low				
Moderate				
High				

B. Earthquake performance of non-structural components

Intensity of shaking	Damage			
	Severe to non-repairable	Repairable: evacuation	Repairable: no evacuation	No significant damage
Low				
Moderate				
High				

C. Continuation of building function: structural and non-structural

Intensity of shaking	Damage			
	Severe to non-repairable	Repairable: evacuation	Repairable: no evacuation	No significant damage
Low				
Moderate				
High				

in the cells of the table that apply if a conventional level of seismic resistance is provided; namely, application of minimum code design standards. It is also helpful if the engineer, with knowledge of local seismicity and code requirements of the region, suggests Modified Mercalli intensity values and average return periods for each of the three levels of shaking. During discussion a client's expectations and requirements can be compared to those normally met by application of the minimum standards proscribed by codes. Clients can consider improving aspects of the seismic performance of their proposed building in order to manage their seismic risk objectives. Many clients will be content with conventional seismic design practice, but others will welcome the opportunity to increase the seismic resilience of their new facility. For some, earthquake insurance or the lack of it may be among the factors to be considered when deciding upon seismic design standards. Perhaps, rather than paying annual insurance premiums, clients may choose to carry the risk of seismic damage themselves and request enhanced seismic performance.

For a client without first-hand experience of a damaging quake, the architect and structural engineer could provide preparatory reading material before discussions begin. If the client has read a realistic earthquake scenario appropriate to the location of the proposed building, all parties approach discussions with a reasonably similar perception of the damaging quake for which the building is designed.[2] There is a reasonable chance that the building and its occupants will be inextricably caught up in and affected by a similar quake.

DESIGN TEAM

The core members of a design team are the architect, the structural engineer and the mechanical services engineer who is usually responsible for mechanical, electrical and plumbing systems. Other specialists, such as fire and façade engineers are co-opted as necessary.

Architects face the temptation to progress a design through to the end of concept stage before involving a structural engineer. While this strategy might succeed for the simplest of buildings, it is generally unwise. Seismic design decisions, such as whether to use shear walls or moment frames, can greatly affect interior planning and exterior elevations. Structural engineering input during this early design phase prevents inappropriate design decisions, minimizes the likelihood of later redesign by the architect, and reduces the client's investment by achieving lower construction costs. If a structural engineer is asked to design a building where spatial planning is well advanced, significant redesign may be required and the structural cost may escalate. It is also probable that the structure, forced into available spaces rather than forming spaces, clashes with rather than complements design concepts.

Architectural education fosters personal initiative, independence of thought and design decision making. So when designing a seismic resistant building an architect must accept that some change in mind-set is required and adapt to working with others, especially the structural engineer as soon as possible. Immediately after an architect has undertaken a *very* preliminary or conceptual structural design that is thoroughly integrated with the design concept and building program, he or she should meet with the structural engineer to refine it.

The importance of early architect-structural engineering interaction cannot be over-emphasized. Architect Christopher Arnold advises:

- Collaboration must occur at the onset of a project: before architectural concepts are developed or very early on in their conception.
- Business conditions that restrict early architect/engineering interaction must be alleviated (by the use of a general consulting retainer fee, for example, recovered from those projects that are achieved).
- If the architect does not want to interact with his [or her] engineer, or if for some reason is prevented by doing so, then he should work with simple regular forms, close to the optimal seismic design.[3]

Not only are all-important and often binding building configuration decisions made during the conceptual stage of a design, but structural advice regarding the width of seismic separation gaps between building and site boundaries are required even before the building plan dimensions are finalized (Chapter 8). There are many other reasons for and advantages of early architect–engineer interaction. For example, foundation conditions might strongly favour one structural system over another, achieving considerable cost savings although impacting upon architectural planning.

At this early stage of a building design many issues with significant architectural and engineering consequences need to be discussed. Table 13.2 facilitates discussion of major points.

Having discussed and resolved the issues raised by Table 13.2, the design tasks can then be apportioned among team members. Take care to ensure that the seismic design of every non-structural component and its connections to the structure is not missed. A designer should be allocated responsibility for each component – for its design through to final site inspection. FEMA 389 also provides a checklist for this process.[1]

CHECKLIST FOR DESIGN TEAM INTERACTION

Item	Minor issue	Significant issue
Goals		
Life safety		
Damage control		
Continued function		
Site characteristics		
Near fault		
Ground failure (landslide, liquefaction)		
Soft soil (amplification, resonance)		
Building configuration		
Architectural concept		
Structural system(s) in each direction		
Vertical discontinuity		
Soft storey		
Setback		
Short column		
Off-set resisting elements		
Plan discontinuity		
Re-entrant corner		
Torsion		
Diaphragm integrity		
Other		
Boundary separation gaps		
Structural systems		
Ductile mechanisms		
Drift/interstorey deflection		
Special systems (e.g. seismic-isolation)		
Repairability		
Non-structural components		
Cladding		
Glazing		
Infills		
Partitions		
Ceilings		
Stairs		
M&E equipment		
Special equipment (computer)		
Other		

CONTRACTOR

Traditionally, architects communicate their design intent to the contractor through plans and specifications. The success of this approach depends on the clarity with which information is presented, the care with which the contractor reads the contract documents and the degree of familiarity the contractor has with the construction detailing as drawn. Where contractors are not familiar with aspects of seismic design and detailing the seismic safety of buildings may be jeopardized.

A US study of construction practice from the perspective of seismic resistance reveals a need to improve communication and collaboration in the construction industry, including between architects and contractors.[4] Using mail surveys and construction site visits, the seismic safety features of wood frame and commercial construction were observed and found lacking.

> 'The results of both surveys show projects with few or no flaws, as well as projects with many flaws. Of the 28 items included in the mail survey, 17 were missing or flawed in over 40 per cent of recorded units. **It is alarming that key items to resist seismic force are among those which are most frequently missing or flawed, including: shear wall hold-downs, nailing, and proportion; wall-to-wall straps and tie-downs; diaphragm blocking and nailing; drag strut splices; roof-to-wall clips and tie-downs.** These items are important links in the force path to transfer seismic or wind forces from roof to foundation'.

Studies like this should be conducted every few years in seismically-prone countries. Unless the quality of construction detailing is independently assessed, architects and engineers cannot be confident of the general quality of seismic resistance or of how well-prepared buildings are for a damaging quake — until it occurs! By then it is too late. Earthquakes are highly effective at exposing design and construction deficiencies. Frequently after destructive earthquakes, calls from within and outside the building industry express the need to improve standards.

By improving their communication with contractors, architects improve the quality of seismic resistance. First, architects can inform. They can explain by way of a report or verbally how the building has been designed to resist an earthquake. Describe the force path so the contractor can understand the reason for those details that seem so strange and unnecessary. Outline the intended ductile mechanism and explain what Capacity Design is and how it has been incorporated into the structural design. Consider revising Table 13.2 to customize it for your project and use it as a checklist. Not only do contractors appreciate being informed and are empowered by this knowledge but they discover like all of us that, from time-to-time some of their long-held assumptions need revision. Seismic design is more sophisticated and the performance of seismic resistant structures more sensitive to construction variations than they realise.

For example, consider the perception that 'stronger is better'. This belief is translated into practice on a construction site when a contractor substitutes a stronger element for a weaker one that is not readily available; such as using larger diameter or higher strength reinforcing bars than those specified. If that additional strength affects a structural fuse region, then brittle fracture will occur somewhere else in the structure, rather than the intended ductile hierarchy. A contractor should understand how a building is intended to perform. As a contractor's knowledge increases, a greater sense of collaboration develops within the construction team.

Another method to improve on-site communication is for the architect, together with the structural engineer, to initiate meetings with the contactor and relevant sub-contractors to discuss less common seismic details. For example, in a region where the separation of infill walls from moment frames is uncommon, a site meeting should be held *before* construction commences to explain the reasons for separation and then discuss how it is to be achieved. If attention is not drawn to new details they may not be constructed properly, if at all. Mistakes and omissions can be corrected but the final result – as well as the attitude of the contractor – will be far more positive if problems are avoided in the first place.

Most aspects of construction require careful attention to achieve good seismic performance. Structural materials should be of high quality and workmanship up to required standards. Weak materials and poor construction are two important reasons why buildings are severely damaged or collapse during earthquakes. An architect should impress upon the client the need for good quality control during construction. A quality control system needs to identify where a contractor deviates from construction drawings and specifications due to genuine mistakes, ignorance or dishonesty. Then remedial action can be taken. Less than rigorous quality control compromises the seismic safety of buildings.

One final comment about the architect–contractor relationship: remember that communication is a two-way process. Although the educative role of the architect has been discussed, architects also learn from contractors. A skilled and experienced contractor has much to share that is of benefit to architects and their projects. In the spirit of collaboration, architects should listen carefully to such wisdom.

POST-EARTHQUAKE

In the aftermath of a destructive quake in an urban area, thousands or maybe hundreds of thousands of buildings require safety assessments. Beginning as soon as possible after the quake the objective of a rapid safety evaluation procedure is to get people back into safe homes and businesses as quickly as possible, and to keep them out of unsafe structures. Building officials, volunteer structural engineers and experienced architects undertake this work. The ideal is for each building in the affected area to receive a quick yet careful assessment. After noting the observed damage on a rapid evaluation assessment form the assessor posts a placard.[5]

If the procedure described in ATC-20 is followed, a green placard indicates that, although damaged, a building is safe for entry and occupancy. If a building poses an imminent threat to safety in all or most of it, a red placard is posted (Fig. 13.1). A 'restricted use' or yellow placard is posted when there is some risk from damage in all or part of the

UNSAFE

DO NOT ENTER OR OCCUPY
(THIS PLACARD IS NOT A DEMOLITION ORDER)

This structure has been inspected, found to be seriously damaged and is unsafe to occupy, as described below:

_____
_____
_____
_____

Do not enter, except as specifically authorized in writing by jurisdiction. Entry may result in death or injury.

Facility Name and Address:

_____
_____

Date _____

Time _____

This facility was inspected under emergency conditions for:

_____
(Jurisdiction)

Inspector ID / Agency

_____
_____
_____

Do Not Remove, Alter, or Cover this Placard until Authorized by Governing Authority

▲ **13.1** Wording on a red placard indicating a damaged building is unsafe to enter (From ATC 20-2).

building that does not warrant red tagging. Entry and occupancy are limited in accordance with restrictions written on the placard.

Different countries have their own rapid evaluation procedures and placards. Architects with, say, five years or more experience are encouraged to attend a training seminar. The local civil defence or emergency management office should be contacted for further information. Apart from making a valuable contribution to a shocked and distraught community, an architect will personally benefit immensely. First-hand observation of seismic damage and communicating its implications to owners and occupants of earthquake damaged buildings is a professional life-changing experience.

REFERENCES AND NOTES

1 Table 13.2 is based upon a similar table in Applied Technology Council (2004). *Primer for Design Professionals: Communicating with owners and managers of new buildings on earthquake risk (FEMA 389)*. Federal Emergency Management Agency, US.

2 A number of earthquake scenarios can be downloaded from the internet, including one very comprehensive example: Stewart, M. (ed.) (2005). *Scenario for a Magnitude 6.7 Earthquake on the Seattle Fault*. Earthquake Engineering Research Institute (EERI) and the Washington Military Department, Emergency Management Division.

3 Arnold, C. (2001). Chapter 6 Architectural considerations. In *The Seismic Design Handbook* (2nd Edn.), Naeim, F. (ed.). Kluwer Academic Publishers.

4 Schierle, G.G. (1993). *Quality Control in Seismic Resistant Construction*. Report to the National Science Foundation on Research, under Grant BCS-9203339. Publisher unknown, Los Angeles.

5 For assessment procedures and forms refer to: AT Council (1989). *Procedures for postearthquake safety evaluation of buildings: ATC- 20*. Applied Technology Council, and (1995). *Addendum to the ATC-20 postearthquake building safety evaluation procedures: ATC-20-2*. Applied Technology Council. ATC placards can be downloaded from www.atcouncil.org/fandp.shtml.

14 NEW TECHNOLOGIES

INTRODUCTION

'New' is a relative term. Compared to the span of architectural history, the whole discipline of earthquake engineering, as outlined in Chapter 3, can be considered new. Only thirty years ago application of the Capacity Design approach enabled structural engineers to design collapse-resistant structures. For the purposes of this chapter, 'new' applies to developments generally less than twenty years old.

One of the exciting features of engaging in such a youthful discipline is the potential for further innovations. Ideas on improving the seismic resistance of buildings are far from being exhausted. With every increase in computing power, introduction of new seismic hardware – like a damper – or after reflection upon the earthquake damage inflicted upon a building, new ideas for seismic resistance surface. After careful analysis many will be discarded, but others may become precursors to or lead directly into important future developments.

Although not experiencing the rapidity of technological change evident in some other fields – such as computing or telecommunications – developments in earthquake resistant design show no sign of abating. Now that researchers and code writers have provided structural engineers with methods to prevent structural collapse in large quakes, at least in developed countries attention is turning to reducing structural and non-structural damage. Non-structural damage costs dominate earthquake repair bills (Chapter 11) even if the potentially huge cost of down-time is excluded. Structural engineers are increasingly aware of the need to reduce these economic impacts and to improve the method by which these vulnerabilities and their solutions are communicated to architects and their clients.

The types of seismic-resistant technologies discussed in this chapter cover a wide range. Beginning with seismic isolation, a simple but revolutionary approach to outwit the quake, the chapter then focuses upon energy absorbing devices, often known as dampers. Next, recent developments in damage avoidance design are considered. After noting the on-going quest for innovative structural configurations notable for their earthquake energy absorbing potential, two proposed design approaches that are currently gathering momentum are introduced. Of the two, performance-based design will be more immediately relevant to architects. Even if too complex and time-consuming on design resources to be requested by many building owners at present, it is certain to impact future code developments. This approach to seismic design, which incorporates a holistic view of seismic performance, aims to achieve greater precision in predicting and limiting both structural and non-structural seismic damage and losses. The chapter concludes by reviewing several other technical developments and future trends in the provision of seismic resistance to buildings.

SEISMIC ISOLATION

The concept of seismic isolation – often called base-isolation – is over one hundred years old. The first patent was filed in 1909 by an English physician who proposed talcum powder as the means of isolating load-bearing walls from their foundations. The application of seismic isolation to real projects did not begin until the late 1970s. Since then, approximately 2000 buildings world-wide have been isolated, mostly in Japan, US, Europe and New Zealand. Japan experienced a dramatic upsurge in seismic isolation following the 1995 Kobe earthquake. Now 1500 buildings are isolated. Of those, and in contrast to statistics from other countries, half are condominiums.[1] With the trauma of Kobe still relatively fresh in their minds, some Japanese are prepared to pay a premium for the increased safety seismic isolation provides. Almost all recently constructed hospitals in Japan have been seismically isolated; as is increasingly the case in the US and New Zealand.

The beauty of seismic isolation is its conceptual simplicity. Imagine a building founded on ball-bearings. During an earthquake the building remains almost stationary while its foundations, subject to the energy of earthquake waves, move violently to-and-fro with the ground. Rolling of the ball-bearings accommodates the relative movement between superstructure and ground (Fig. 14.1). Vertical isolation is unnecessary given the generally excellent performance of buildings to vertical shaking.

Although the concept is simple, its implementation is more complex. If the interface between superstructure and foundations is almost frictionless, what happens during a wind storm? Although TV reporters would flock to a building being blown along a street, its occupants would be less impressed. Any seismic isolation system must be wind resistant. It should also possess a centring mechanism to continuously force the isolated superstructure to return to its original position during

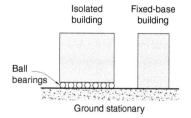

Isolated building   Fixed-base building

Ball bearings

Ground stationary

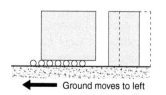

← Ground moves to left

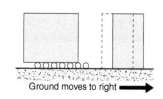

Ground moves to right →

▲ 14.1   Relative movement between the ground and a perfectly isolated building on ball bearings, and a fixed-base (conventional) building.

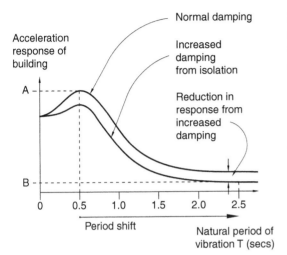

Normal damping

Increased damping from isolation

Reduction in response from increased damping

Acceleration response of building

A

B

0   0.5   1.0   1.5   2.0   2.5

Period shift

Natural period of vibration T (secs)

▲ **14.2** A response spectrum showing how the peak acceleration of a fixed-base building, A, is reduced to B by the 'period shift' and increased damping of a seismic isolation system.

shaking. Finally, damping is necessary. It reduces the seismic response of the superstructure and reduces the relative horizontal deflections that bearings and other details need be designed for.

Using terminology introduced in Chapter 3, seismic isolation reduces the seismic response of a super-structure by increasing its natural period of vibration and increasing its damping. These two interventions are shown in Fig. 14.2 which illustrates the effect of isolating a typically-damped conventional fixed-base building with a natural period of 0.5 seconds. During a design-level earthquake its peak acceleration is given by point A. The provision of a horizontally flexible iso-lation interface increases its natural period to, say, 2.5 seconds. At this far longer period of vibration the seis-mic response is considerably less, and reduced even further by the additional damping provided between superstructure and foundations. Point B denotes the peak acceleration of the isolated building, typically 20 per cent to 25 per cent of A.

The ability of seismic isolation to reduce seismic response or peak accelerations of buildings has been confirmed by observation and vibration measurements during several earthquakes. During the 1994 Northridge, Los Angeles, earthquake the seismically isolated University of Southern California Teaching Hospital came through unscathed while the nine nearby hospitals were so badly damaged all had to be evacu-ated. The following account summarizes acceleration measurements for two buildings in separate quakes; the university hospital above, and a computer centre in Japan during the 1995 Kobe earthquake:[2]

> The measured free field peak ground acceleration … was 0.49 g, while the peak acceleration throughout most of the structure was less than 0.13 g, and the peak acceleration at the roof was 0.21 g due to structural amplification in the upper two storeys. By analys-ing a model of the hospital without isolation, Asher et al. concluded that accelerations throughout the fixed base structure would have ranged between 0.37 g and 1.03 g, and that damage to building con-tents and disruption of service would have been almost certain.
> The West Japan Postal Savings Computer Center experienced ground motions with a peak site acceleration of 0.4 g in the January 17, 1995 Kobe earthquake (DIS1996). The Computer Center exhib-ited no damage, and the maximum recorded acceleration in the building was 0.12 g. A nearby fixed base building of approximately

the same height was also instrumented, and the maximum recorded acceleration at the roof was 1.18g.

Seismic isolation of the seven-storey university hospital reduced its peak ground acceleration to approximately half that of the ground. But as Bill Robinson points out, seismic isolation is more effective than that:

> This 7-storey hospital underwent ground accelerations of 0.49g, while the rooftop acceleration was only 0.21g – a reduction by a factor of 1.8. The Olive View Hospital, nearer to the epicentre of the earthquake, underwent a top floor acceleration of 2.31g compared with its base acceleration of 0.82g, a magnification by a factor of 2.8. A comparison between the hospital seismically isolated with lead-rubber bearings, the University Teaching Hospital, and the unisolated building, the Olive View Hospital, shows an advantage by a factor of 1.8 × 2.8~5 in favour of the isolated hospital.[3]

The performance of the Japanese computer centre building shows an even larger reduction of peak acceleration. Not only does the superstructure of a seismically isolated building experience less peak acceleration than its foundations but, unlike a fixed-base building, it does not amplify accelerations significantly up its height.

Similarly impressive reductions in superstructure accelerations were measured in a building during two lesser recent Japanese earthquakes.[1] Not only do the occupants and contents of a seismically isolated building experience far less severe accelerations, any motion is felt as gentle back-and-forth or side-to-side movements with a period of vibration of between 2.0 to 3.0 seconds. This is a very gentle ride compared to the intense flinging and whipping motions within conventional buildings. In isolated buildings almost all of the relative movement between foundations and roof occurs at the plane of isolation. The superstructure tends to move as a rigid body in stark contrast to a fixed-base building where significant inter-storey drifts occur. No wonder seismic-isolation is an attractive option for buildings required to be functional immediately after a damaging quake, or housing expensive or irreplaceable contents.

Although conceptually compelling and with an impressive, albeit limited track-record, seismic isolation is definitely not a panacea for all seismic ills. The following considerations limit its applicability:

- Flexible buildings, generally more than ten-storeys high with natural periods of vibration greater than 1.0 seconds may not benefit sufficiently from the 'period shift' to between 2.0 and 3.0 seconds.
- Sites underlain by deep soft soils have their own long natural periods of vibration with which an isolated system with a similar natural period could resonant.

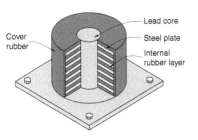

▲ **14.3** A cut-away view of a lead-rubber bearing.

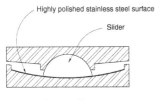

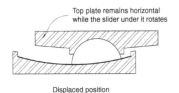

▲ **14.4** A section through a Friction Pendulum™ bearing.

▲ **14.5** A Friction Pendulum™ bearing at the base of a column. The temporary locking brackets are yet to be removed.
(Reproduced with permission. © Skidmore, Owings & Merrill, LLP).

- Relatively wide horizontal separation gaps to allow for relative movement between superstructure and foundations in the order of 400 mm may represent a serious loss of usable floor area for buildings on tight urban sites.
- The first cost of a seismically isolated building is generally a few percent more than that of a fixed-base building. But, as Ron Mayes points out, if a client decides against earthquake insurance, income from saved annual premiums that is invested can repay the cost of isolation within three to seven years. When a damaging quake occurs possibly uninsurable business disruption costs 'will overwhelm any first cost consideration especially if the building contents have any significant value'.[4]
- Some current codes still take a very conservative attitude to this system that has been rarely tested in real buildings. They limit its potential advantages and dampen designers' enthusiasm by what some consider overly onerous analytical and testing procedures. Fortunately this situation is improving.

A seismically isolated building still requires vertical structures such as shear walls or moment frames. Like any other building, wind forces need to be resisted and even though isolated, seismic forces require adequate force paths. But, whereas considerable damage is expected to the ductile structure of a fixed-base building in a design-level earthquake, the structure of an isolated building would normally remain undamaged. Seismic isolation also allows buildings with normally unacceptable configurations (Chapters 8 and 9) or with absolute minimum vertical structure to perform adequately.

The horizontal flexibility and centring capability of a seismic isolation system is usually achieved by a sliding or shearing type of mechanism. The controlled rocking capability of slender structures has also been utilized in several structures, including in the Hermès building mentioned in Chapter 3. A range of special structural components facilitate these mechanisms. Lead-rubber bearings, with internal steel plates to limit vertical deflections caused by forces from the columns or walls they support, are popular (Fig. 14.3). As the top of a bearing is displaced sideways the elasticity of the rubber provides an elastic restoring force. Damping is provided by an internal lead (Pb) plug, or by another damping device. Sometimes the rubber is formulated with inherently high damping characteristics. Friction pendulum dampers, like those used in the San Francisco International Airport, require less vertical space (Figs 14.4 and 14.5). The gentle

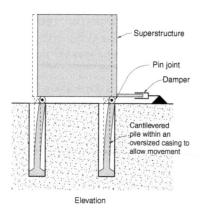

Elevation

▲ **14.6** Schematic representation of an isolation system where cantilevered piles provide horizontal flexibility and centring forces.

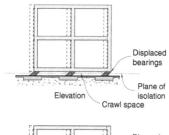

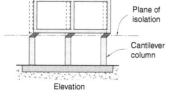

▲ **14.7** Two possible locations for an isolation plane.

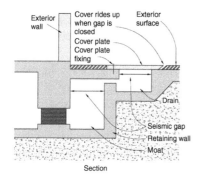

Section

▲ **14.9** Possible section through the perimeter of a seismically isolated building.

▲ **14.8** Lead-rubber bearings on cantilever reinforced concrete columns. Bhuj Hospital, India
(Reproduced with permission from Adam Thornton).

curved dish, which causes the column above to slide back down towards the lowest point, provides the centring mechanism. Two buildings in New Zealand are founded on long piles separated from the surrounding ground to achieve the long natural period necessary for seismic isolation (Fig. 14.6). A structural engineer should discuss the advantages and disadvantages of each isolation system with the architect before a final choice is made.

When designing a seismically isolated building, an architect must collaborate with the structural engineer over a number of unique design decisions. The most fundamental is the location of the plane of isolation. Should the isolators be placed in a crawl-space under the lowest floor level or can they be located at the top of cantilever ground floor columns (Figs 14.7 and 14.8)? If the first configuration is chosen, a 'moat' to accommodate a wide separation gap is required together with a suitable cover. The moat cover is designed to offer insignificant restraint to relative movement across or along the gap (Fig. 14.9). For an elevated isolation plane the design challenge involves detailing some

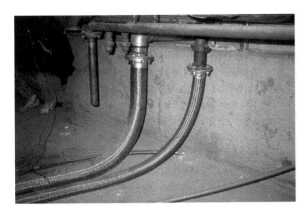

▲ **14.10** Flexible services pipes. Seismically isolated office building, Wellington.

▲ **14.11** San Francisco City Hall.

▲ **14.12** Parliament Buildings, Wellington.

vertical architectural elements, like the cladding that crosses the plane of isolation, to allow horizontal movement without restraint or damage.

Services connections to an isolated building must also accommodate relative movement. Gas, water and sewer pipes, as well as electrical and communications wiring require the capacity to undergo design movements without rupturing (Fig. 14.10). Even elevator shafts and their pits that project beneath ground floors need to be separated from the ground. *Nothing* must compromise unrestrained horizontal movement between the isolated superstructure and foundations.

Seismic isolation is a viable option for both new buildings and those requiring seismic upgrading or retrofitting. Notably in the US and New Zealand, a number of load-bearing masonry buildings, including some of historical and cultural importance, have been seismically isolated (Figs 14.11 and 14.12)[5,6] In many situations, including the previous two examples, seismic isolation alone is insufficient to guarantee adequate retrofitted seismic performance. New structural elements, like braced frames or shear walls, are also required. Seismic isolation of existing masonry buildings requires significant construction effort in the foundation area. Every wall and column of the existing building needs to be sliced through to form the plane of isolation. After upgrading existing foundations and bases of the walls, bearings are inserted. They support the entire weight of the building while simultaneously providing horizontal flexibility, and often damping as well.

Several modern buildings have also been retrofitted using seismic isolation. When the Victoria University of Wellington discovered that, even after a moderate to major quake, their library building would be severely damaged, books spoilt and the library inoperable for several years, a decision was made to improve the building's seismic resistance

▲ **14.13** An area of the podium is propped while excavation is completed prior to installation of lead-rubber bearings. Library, Wellington.

▲ **14.14** A lead-rubber bearing inserted at the base of a reinforced concrete column.

Moment
frame
Damper
Diagonal
brace

Section

▲ **14.15** Dampers at each level of a moment frame to reduce accelerations and drifts.

(Fig. 14.13). In what is best described as sophisticated seismic surgery, each of sixteen tower-block columns in turn had their gravity forces supported on temporary steel props and hydraulic jacks. After two cuts with an abrasive wire a concrete block at the base of each column was removed and a lead-rubber bearing inserted (Fig. 14.14). A moat constructed around three sides of the building basement allows for horizontal movement. It is covered by a sliding plate.[7] The whole operation was completed without disrupting library operations. The only disappointing feature of the project is how the architectural detailing for seismic isolation denies both the radical surgery undertaken and how the library is now flexibly attached to its foundations.

In summary, seismic isolation is conceptually the best method for outwitting a quake. Although there are situations for which isolation is unsuitable, where it is applied most of the earthquake energy that would otherwise cause extensive damage to the building fabric and its contents is prevented from entering the building. This is the closest we come to 'earthquake-proof' construction. Isolated buildings ride through intense and violent shaking of major quakes with no more than sedate and gentle movement.

DAMPERS

Dampers perform the same function in a building as shock absorbers do in a motor vehicle. They absorb vibration energy in the system, reducing acceleration and movement to provide a smoother ride. Dampers are also known generically as energy dissipaters. They transform dynamic energy into heat, either reducing horizontal drifts in buildings (and therefore damage), or allowing designers to specify more slender structural members. The role of dampers in seismic isolation where they have most often been used to date has already been mentioned. But dampers can also be placed within superstructure frameworks; perhaps as diagonal members within a new or even an existing moment frame to absorb energy and reduce inter-storey drifts (Fig. 14.15).

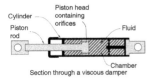

Cylinder
Piston head containing orifices
Piston rod
Fluid
Chamber
Section through a viscous damper

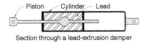

Piston — Cylinder — Lead
Section through a lead-extrusion damper

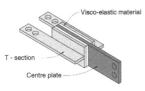

Visco-elastic material
T - section
Centre plate
Visco-elastic damper

Rod pushes and pulls cantilever
Direction of movement
Mild steel cantilever
Steel cantilever damper

▲ **14.16** Four types of damping devices.

▲ **14.17** Dampers placed at the ends of tubular steel braces. Office building, San Francisco.
(Reproduced with permission from David Friedman, R. Cranfield, photographer).

▲ **14.18** A lead-extrusion damper bolted to the foundations (left) and to the isolated superstructure (right). Police Station, Wellington.

The shapes, materials and methods for absorbing energy vary greatly among dampers (Fig. 14.16). In a classical viscous damper, like those sometimes found in college physics labs, a piston with an orifice is forced through a fluid-filled cylinder (Fig. 14.17). The resistance of the damper is proportional to the velocity of the piston. In a lead-extrusion damper, the cylinder is filled with lead instead of fluid. The movement of its bulb is minimal until the resistance of the lead peaks. Then the pressure of the steel bulb causes the lead to deform plastically, allowing the bulb to move within the enclosing steel cylinder (Fig. 14.18). A thin film of visco-elastic material, which absorbs energy when subject to shearing actions, is the basis of another type of damper. If the visco-elastic material is removed and the layers of metal are clamped by bolts in slotted holes, one has the beginnings of a friction damper. The ability of mild steel, whether deformed in bending or torsion leads to yet another set of damping devices, possibly the most simple. Dampers utilizing shape memory alloys that absorb energy, yet return to their original shape after being deformed, are currently under development but are yet to be installed in a seismic resisting system.

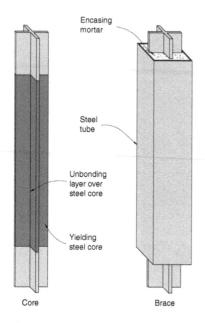

Encasing mortar

Steel tube

Unbonding layer over steel core

Yielding steel core

Core

Brace

▲ **14.19** Schematic details of a buckling-restrained brace.

▲ **14.20** Buckling-restrained braces improve the seismic resistance of an existing building. Berkeley, California.

A buckling-restrained brace is another structural device able to dissipate energy (Figs 14.19 and 14.20). It consists of an inner steel member capable of resisting tension or compression forces that also functions as a structural fuse – yielding and absorbing energy. An outer tube and a grout-filled gap between outer and inner members prevent the inner member from buckling, and ensure a highly ductile and dependable performance. Depending on the strength of a brace it can be considered either a damper, or if very strong, as a diagonal member of a braced frame. The design of buckling-restrained braces prevents any chance of their buckling and minimizes their damage when deformed into the plastic range. They are preferable to concentrically braced steel frames whose diagonals are prone to buckling. They are also an alternative to eccentrically braced frames whose damaged fuse regions would be difficult to repair after a major quake.

The dampers discussed above are classed as *passive*, in contrast to *active* systems which rely upon computer-controlled hydraulic rams or other devices. Conceptually, active systems represent an exciting advance over passive devices. The idea of structural members or devices responding to signals from sensors so as to reduce or even eliminate seismic energy input into a building is most attractive, particular to those people drawn to high-tech solutions. Unfortunately, although conceptually simple, the practical application of active control is still some years away and even then is likely to be very expensive.

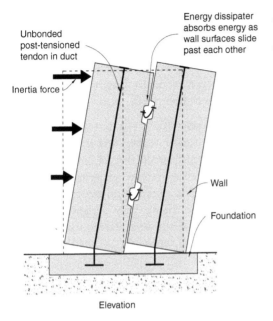

Unbonded post-tensioned tendon in duct

Energy dissipater absorbs energy as wall surfaces slide past each other

Inertia force

Wall

Foundation

Elevation

▲ **14.21** Damage avoidance precast concrete shear wall system.

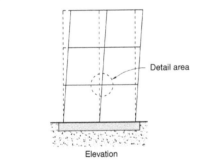

Detail area

Elevation

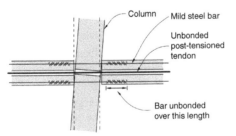

Column

Mild steel bar

Unbonded post-tensioned tendon

Bar unbonded over this length

Elevation of central joint when frame drifts sideways due to seismic forces

▲ **14.22** Damage avoidance detailling in a precast concrete moment frame beam-column joint.

Although in 1998 more than twenty buildings in Japan had active computer-controlled systems to reduce wind vibrations, only four systems continue to work during an earthquake. Akira Nishitani sums up the state-of-the-art in Japan:

> … *active control still has a long way to go to achieve the highest goal: to have reliable active control systems which could ensure structural safety even during strong earthquake. To do this, several problems have to be solved, eg. how to supply the external power which is needed or how to make systems with less reliance on external power, and how to increase the reliability and robustness of active control systems.*[8]

DAMAGE AVOIDANCE

Given that the best method of damage avoidance, seismic isolation, is unsuitable in many situations, researchers are investigating other methods of avoiding damage – particularly to structural members. To a large degree, buckling-restrained braces achieve this goal but chunky diagonal braces are often unacceptable architecturally. Researchers have proposed several methods whereby damaged structural fuses are easily replaceable but none have been widely adopted.[9]

Precast concrete industry-sponsored research in the US has developed damage avoidance systems for precast concrete moment frames and shear walls (Figs 14.21 and 14.22).[10] The two essential elements of these systems are an internal centring spring action and energy dissipation by either easily replaceable fuses or fuses that remain undamaged. Unbonded (ungrouted) post-tensioned tendons, which allow relative rotational movement between components like beams and columns, yet spring them back to their original positions, are central to both systems. The spring-back behaviour is just like that of children's toys which consist of wooden pieces joined by internal lengths of elastic thread. After distorting the toy, the elastic thread snaps the toy back to its original geometry. Energy dissipation is provided by mild steel bars or other ductile devices that yield when deformed in shear, as in vertical gaps up shear walls, or elongate in tension and squash in compression where placed within beams of moment frames. While the behaviour of these damage-free vertical systems themselves are excellent,

▲ **14.23** Post-tensioned unbonded tendons provide a damage avoidance mechanism in most of the moment frames of this 39-storey apartment and retail building, San Francisco.

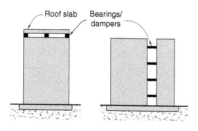

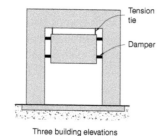

Three building elevations

▲ **14.24** Theoretical examples of where discontinuities and dampers are inserted into buildings with the intention of reducing the overall seismic response.

structural engineers need to pay close attention to how floor slabs affect their seismic performance. While it is desirable from the perspective of damage avoidance to separate slabs from the beams of moment frames this diminishes the essential role that floor slabs play as floor diaphragms to effectively tie a building together at each floor level.

From an architectural perspective these precast concrete structural systems are very similar in form and size to monolithic reinforced concrete construction (Fig. 14.23). Both precast and cast-in-place concrete systems experience similar seismic drifts and accelerations causing non-structural damage and structural damage to the specially designed precast concrete members during a design-level earthquake. This is expected to be minor and easily repairable.

INNOVATIVE STRUCTURAL CONFIGURATIONS

The search for innovative structural configurations to outwit quakes is still very much alive, after having exercised many minds over the centuries. Indeed, the list of those searching has included more than structural engineers. Witness a physician's patent for base-isolation, and architect Frank Lloyd Wright's commendable yet flawed attempt to use soft ground as a seismic isolation system for the Tokyo Imperial Hotel.[11] The discovery of the ultimate 'earthquake-proof' system or configuration remains tantalizingly elusive.

Researchers and practicing engineers share their ideas and even brainstorm in professional engineering journals and technical conferences. While acknowledging that they are still journeying towards the goal, they keep the quest alive, hoping their ideas might be developed by or inspire others. As an example, preliminary analyses suggest that a seismically isolated concrete roof slab designed to vibrate out-of-phase with the multi-storey structure beneath has the potential to reduce the amount of vertical seismic structures.[12] Other researchers have proposed deliberately introducing discontinuities into buildings and then connecting the various sections by energy absorbing devices to reduce overall levels of seismic force (Fig. 14.24).[13] In two separate papers, Eric Elsesser illustrates the historical development of US West Coast multi-storey seismic framing systems. Then, in the spirit of encouraging innovation, suggests possible configurations and structural systems with excellent energy absorbing characteristics.[14,15] All these ideas invite further analysis and testing. During that process one or two may be applicable to particular projects. But most, like the fallen seeds of trees and plants, lie dormant. Who knows if and when some day they might take root and bear fruit?

STRUCTURAL DESIGN APPROACHES

Two new approaches to structural design are gathering momentum within the structural engineering profession. Of the two, 'displacement-based design' has fewer implications for architects although it does provide a more rational basis for achieving acceptable levels of performance in design earthquakes. Rather than a structural engineer commencing a seismic design by calculating the stiffness and natural period of a building and then determining inertia forces, displacement-based design begins by specifying a maximum acceptable displacement and then determines the forces that condition causes.[16] Displacement-based design will probably eventually replace the current method of force-based design.

Performance-based Seismic Design, which has considerably more ramifications for architects and their clients, is the second new design approach. FEMA 349 offers an explanation:

> The concept of Performance-based Seismic Design (PBSD) is to provide engineers with the capability to design buildings that have a predictable and reliable performance and permit owners to quantify the expected risks to their buildings and select a level of performance that meets their needs while maintaining a basic level of safety. PBSD uses the concept of objectives, allowing the owner to specify an acceptable level of damage to a building if it experiences an earthquake of a given severity. This creates a 'sliding scale' whereby a building can be designed to perform in a manner that meets the owner's economic and safety goals.[17]

Performance-based Seismic Design, which takes a far more holistic view of seismic design than most current approaches that concentrate on life-safety performance, is currently under development.[18] When applied to a particular building, the method begins by subjecting the building computer model to a series of previously recorded earthquake events approximating a design-level earthquake. Then, by utilizing a database that relates structural and non-structural damage and repair costs to peak drifts and accelerations, the total repair costs incurred can be determined. Eventually, losses such as incurred by down-time and casualties may be calculable.

With such a sophisticated and powerful tool to model damage and down-time costs, architects and engineers will be able to quantify many of the issues that clients need to be informed of and make decisions about (Chapter 13). It offers potential for far more detailed explorations of design alternatives with respect to the cost implications of their seismic performance. Even if this design approach is not used in structural engineers' design offices on a day-to-day basis due to its current complexity, it will be useful in comparing the performance of different structural systems and will inevitably influence seismic code developments.

Performance-based Seismic Design concepts have been applied to improving the seismic resilience of the University of California Berkeley campus. A seismic performance goal was to limit a post-earthquake campus closure to thirty days. With reference to one major laboratory building, Mary Comerio outlines a very detailed assessment of the vulnerability of lab equipment and other building contents. After computer modelling the building and shake-table tests of selected lab contents, equipment damage was determined and measures to mitigate it and consequent down-time were taken.[19]

OTHER DEVELOPMENTS

Seismic resistant design options increase with the development and introduction of new construction materials. In recent years fibre reinforced composite materials have entered mainstream seismic engineering. Carbon fibres, approximately ten times stronger than mild steel of the same cross-section, are combined with a synthetic resin in structural applications. Due to the high cost of composites they are used sparingly and mainly for seismic upgrading. They have been used to confine poorly reinforced columns by circumferential wrapping or bandaging, and to resist diagonal tension forces within unreinforced masonry walls (Fig. 14.25).

▲ **14.25** A carbon fibre strip glued onto an unreinforced masonry wall to increase its shear strength. Hotel, Wellington.

Shape memory alloys have the potential to improve the effectiveness of dampers. It is likely the properties that allow them to be stretched or compressed and absorb energy, yet return to their original position, will be exploited for the sake of improved energy dissipation. Steel and other types of short fibres added to concrete enable reduction in structural member sizes; as does the introduction of self-compacting and ultra high-strength concrete. Many of these developments widen architecture-enhancing structural options such as increased structural slenderness.

Continued developments in computer hardware and software also benefit earthquake engineers, and therefore architects. Complex structural configurations, which once would have been impossible to design safely with confidence – such as the CCTV building, Beijing (Chapter 6) – are now achievable. Formerly, such geometrically complex buildings would have been simplified at preliminary design stage and seismically separated into independent structures in order for them to be modelled and analysed. Future technological advances can only continue to provide new options and solutions to the problems of improving the seismic performance of buildings. Architects and their clients will be first among the beneficiaries.

REFERENCES

1 Kani, N., Takayama, M. and Wada, A. (2006). Performance of seismically isolated buildings in Japan: observed records and vibration perception by people in buildings with seismic isolation. *Proceedings of the 8th US National Conference on Earthquake Engineering, April 18–22, San Francisco, California.* Paper no. 2181, 10 pp.

2 Taylor, A.W. and Igusa, T. (eds.) (2004). *Primer on seismic isolation.* American Society of Civil Engineers.

3 Robinson, W. *Congratulations.* www.robinsonseismic.com (24/02/2006)

4 Mayes, R.L. (2006). Implementation of higher performance structural systems in the United States. *Proceedings of the 8th US National Conference on Earthquake Engineering, April 18–22, San Francisco, California.* Paper no. 2180, 10 pp.

5 Buckle, I.G. (1995). Application of passive control systems to the seismic retrofit of historical buildings in the United States of America. *Construction for Earthquake Hazard Mitigation: Seismic Isolation Retrofit of Structures: Proceedings of the 1995 Annual Seminar of the ASCE Metropolitan Section; New York, February 6–7,* American Society of Civil Engineers, 91–97.

6 Poole, R.A. and Clendon, J.E. (1992). NZ Parliament buildings seismic protection by base isolation. *Bulletin of the New Zealand National Society for Earthquake Engineering,* 25:3, 147–160.

7 Clark, W.D. and Mason, J.E. (2005). Base isolation of an existing 10-storey building to enhance earthquake resistance. *Bulletin of the New Zealand Society for Earthquake Engineering,* 38:1, 33–40.

8 Nishitani, A. (1998). Application of active structural control in Japan. *Progress in Structural Engineering and Materials,* 1:3, 301–307.

9 For example, Dutta, A. and Mander, J.B. (2006). Seismic design for damage control and repairability. *Proceedings of the 8th US National Conference on Earthquake Engineering, April 18–22, San Francisco, California.* Paper no. 2178, 10 pp.

10 Priestly, M. J. et al (1999). Preliminary results and conclusions from the PRESSS five-story precast concrete test building. *PCI Journal,* November–December, 42–67.

11 Wright, F.L. (1977). *Frank Lloyd Wright: an autobiography,* Quartet Books, p. 237.

12 Villaverde, R. and Mosqueda, G. (1998). Architectural detail to increase the seismic resistance of buildings. *Structural Engineering World Wide*, Elsevier Science, Paper T204–2, 8 pp.

13 Mezzi, M., Parducci, A. and Verducci, P. (2004). *Proceedings 13th World Conference on Earthquake Engineering*, Vancouver, BC, Canada. Paper no. 1318, 12 pp.

14 Elsesser, E. (2006). Past, present and future of seismic structures. *Proceedings of the 8th US National Conference on Earthquake Engineering*, April 18–22, San Francisco, California. Paper no. 2179, 10 pp.

15 Elsesser, E. (2006). New ideas for structural configurations. *Proceedings of the 8th US National Conference on Earthquake Engineering*, April 18–22, San Francisco, California. Paper no. 2182, 9 pp.

16 Priestley, M.J.N., Calvi, M.C. and Kowalsky, M.J. (2007). *Displacement-Based Seismic Design of Structures*. IUSS Press, Pavia.

17 EERI (2000). *Action plan for performance based seismic design (FEMA-349)*. Federal Emergency Management Agency.

18 ATC (2006). *Guidelines for seismic performance assessment of buildings (ATC-58 25 per cent draft)*. Applied Technology Council, Redwood City, California.

19 Comerio, M. (2006). Performance engineering and disaster recovery: limiting downtime. *Proceedings of the 8th US National Conference on Earthquake Engineering*, April 18–22, San Francisco, California. Paper no. 531, 10 pp.

15   URBAN PLANNING

INTRODUCTION

Whereas previous chapters have focused upon the seismic resistance of individual buildings, this chapter takes a broader perspective. It discusses how urban planning can reduce a quake's destructive impact upon a region, city or community.

By implementing the principles already outlined in this book, the built environment (including houses, offices, schools and hospitals) has a far better chance of surviving the ravages of a damaging quake. Seismic resistant design enables buildings to withstand low-intensity shaking without damage and higher intensity shaking without collapse. Similarly, seismically-aware urban planning reduces societal vulnerability. Just as public health initiatives, like provision of potable water and sanitation, prevent widespread disease, seismically-aware urban planning has communities build in safer areas that are less vulnerable to strong shaking and other hazards. Ideally, urban planning also enables robust infrastructures to escape immediate post-earthquake paralysis and thereby facilitates the earthquake recovery process.

Urban planning literature highlights the complexities and limitations of planning processes. Planners respond to many diverse and often competing interests that seek to drive planning processes and outcomes. Among these interests, although hardly evident from the reading of urban planning texts, should be a concern for the provision of an adequate level of societal seismic resilience. Often lacking influential support, planners must take whatever actions they can to mitigate seismic impacts upon their society. This chapter shows how urban planning can contribute to the public good by improving seismic safety.

Architects' relationships to urban planning generally take one of two forms. On one hand, a few architects are actively involved as members of the planning profession drawing upon their architectural training and skills. At the other extreme, most architects are subject to urban planning regulations. Architects comply with the requirements of regional

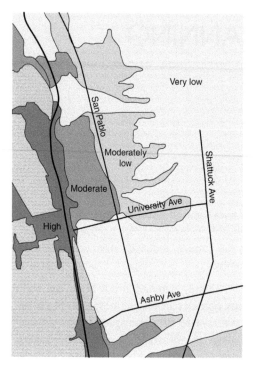

Very low

San Pablo

Moderately
low

Shattuck Ave

Moderate

University Ave

High

Ashby Ave

▲ 15.1   Liquefaction hazard levels are shown in this liquefaction hazard map for Berkeley/Albany, California for an earthquake of the same magnitude as the 1906 San Francisco earthquake.
(Adapted from an ABAG earthquake map).

and city plans as they design for their clients. Irrespective of an architect's relationship to urban planning, planning issues including those related to seismic safety need to be addressed.

PLANNING

Coburn and Spence outline how planners in seismically active regions can integrate earthquake protection into their planning processes.[1] They suggest that planners add microzone and vulnerability maps to their planning tool boxes. Microzone maps, perhaps better known as seismic hazard maps, indicate areas likely to experience more intense shaking than others, perhaps due to deposits of deep soft soil. Areas prone to liquefaction or slope instability above certain earthquake intensity levels, and areas likely to be affected by surface faulting, are also indicated on seismic hazard maps. A State of California seismic hazard zone map shows that significant areas of the cities of San Francisco, Berkeley and Oakland are prone to liquefaction and earthquake-induced landslides (Fig. 15.1).[2] The map clearly explains its aims:

This map will assist cities and counties in fulfilling their responsibilities for protecting the public from the effects of earthquake triggered ground failure as required by the Seismic Hazards Mapping Act.

Other maps of the same region provide additional hazard information in the form of peak ground acceleration contours for earthquakes with certain return periods; or they show the position of active fault lines. While some cities in quake-prone areas have been microzoned, that process needs to be undertaken in all locations so that both planners and society can benefit from this knowledge of seismic hazard.

Seismic hazard information is vital for both proposed and existing built environments. Plans for new developments benefit from land-use information which incorporates the likelihood of severe ground shaking and deformation. Those facilities most important to a community after an earthquake, such as hospitals and fire stations, should be located in the safest areas; or if not, designed so that they will remain operable. Their accessibility should not be compromised by the loss of lifelines like roads and bridges that could be damaged, particularly if built on soils prone to liquefaction or some other earthquake-induced soils failure. It is unwise to locate hazardous operations – like oil or gas storage facilities – on soft soil sites unless the sites are improved beforehand (Chapter 7). Particularly vulnerable areas of land should be designated as open recreational spaces. Application of seismic hazard mapping information improves public safety.

When applied to existing cities and suburbs, seismic hazard maps guide efforts to reduce seismic vulnerability incrementally. Examples include:

- Planning to relocate an essential post-earthquake facility, like a hospital, to a safer site when the time comes to replace it
- Re-routing a life-line such as a water main or major road to avoid an area likely to liquefy or settle, thereby avoiding severe disruption
- Impose development restrictions on seismically hazardous areas of the city while encouraging development in other areas, and
- Rather than redevelop and intensify occupancy in a decrepit area of a city threatened by unstable slopes, turn it into a park to reduce vulnerability to earthquake-induced landslides.

Other earthquake safety enhancements can be undertaken with the assistance of a seismic vulnerability map. This document would show the relative seismic vulnerability of the building stock in a given area based on building surveys and engineering analysis. Built-up areas are classified according to their vulnerability based upon various factors including the ages of buildings, their condition and materials of construction, the seismic design standards applicable at the time of construction, and typical configuration defects. When overlaid on a seismic hazard map, the likely geographical distribution of seismic damage can inform the planning process. Some examples of how city authorities might use this information to reduce seismic vulnerability include:

- Require and assist owners of seismically vulnerable buildings to upgrade the seismic performance of their buildings to protect a specific precinct of historical importance
- Pre-plan the post-earthquake redevelopment of a particularly vulnerable area of the city in the expectation that the next damaging quake will provide a unique window of opportunity[3]
- As opportunities arise, purchase and demolish pockets of old and vulnerable buildings to create multi-purpose open spaces which are part of a coherent open space plan for the city. These are areas where shocked survivors can congregate and find shelter and possibly take refuge from post-earthquake fires, and perhaps where post-earthquake rubble can be piled temporarily, and
- Acquire rows of properties in the most vulnerable areas to increase street widths. As well as reducing day-to-day congestion and enhancing access by emergency services vehicles when streets are strewn with rubble from adjacent buildings, another benefit is the provision of wider fire breaks in anticipation of post-earthquake fires.

Hossein Bahrainry describes how various planning approaches, including some of those outlined above, are applicable to the seismically vulnerable city of Rasht in Northern Iran.[4] The seismic prevention master plans for the cities of Tehran and Istanbul are other examples of these strategies.[5,6] In Peru, the Sustainable Cities Program was officially inaugurated in 1998. Its primary aim is for urban expansion to occur in low-hazard zones.[7]

Urban planners are also able to reduce seismic damage to lifelines. Use of seismic hazard maps in urban planning, with input from lifeline engineering specialists, improves the quality of site selection for lifeline components such as dams, wastewater treatment plants and transportation systems.[8]

Surface fault rupture

One of the most obvious steps that planners take to reduce the seismic vulnerability of a community is to limit construction over known active fault lines. Any building subject to fault movement beneath its foundations can expect severe damage depending upon the magnitude of fault displacement. Movement along a fault line may be confined to the horizontal plane inducing a shearing-action into the foundations of a building above it. Yet perhaps significant vertical movement results in part of a building attempting to cantilever over the lowered ground. The distortions induced in a building straddling a fault cause far more severe damage than that caused by ground shaking alone. Because buildings sited over a ruptured fault are highly likely to at least partially collapse, the increase in the number of casualties can be significant. Fortunately, in even the most fault-ridden cities, the percentage of affected sites is very small due to the narrowness of surface fault zones which are usually of the order of several tens of metres wide.

The Californian Alquist-Priolo Earthquake Fault Zoning Act of 1972 requires the State Geologist to 'delineate "Earthquake Fault Zones" among known active faults...'[9] Local government agencies of affected areas require geological investigation of the hazard. Any building for human occupancy must be set back 15 m from an active fault. Single-family light-weight dwellings up to two storeys high are among several exemptions. While adopting a similar approach, recent New Zealand guidelines take the Californian requirements a step further by taking into account factors such as the accuracy with which a fault can be located, its average faulting recurrence interval and the importance of the proposed building to the community (Fig. 15.2).[10]

▲ **15.2** A suburb is planned so that an active fault (white dashed line) passes through a green strip. Upper Hutt, New Zealand.
(Lloyd Homer, Institute of Geological & Nuclear Sciences, Ltd, New Zealand).

▲ 15.3 Devastation to the built environment by the 2004 Indian Ocean tsunami. Banda Aceh, Indonesia.
(Reproduced with permission from Regan Potangoroa).

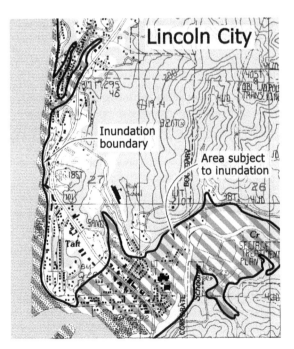

▲ 15.4 An inundation map for a section of the US West Coast for a tsunami caused by a magnitude 8.8 undersea earthquake. (Adapted from a map, State of Oregon, Dept. of Geology and Mineral Industries).

The need for such guidelines to reduce hazards from surface faulting has been dramatically emphasized by the July 2007 earthquake damage to the Kashiwazaki nuclear power plant in Japan. Apparently the plant is built directly above an active fault line: 'The long, straight ridge and crevice that now runs alarmingly through the middle of the plant, said one of Japan's most respected seismologists, proves that'.[11]

TSUNAMI

The devastating 26 December 2004 Sumatra, Indonesia, earthquake and the Indian Ocean tsunami have intensified awareness of this oceanic seismic hazard (Fig. 15.3). Large stretches of coastline around the Pacific Rim and elsewhere are at risk of tsunami attack. The destruction and loss of life from tsunamis is well documented in the histories of numerous villages and cities worldwide. Edward Bryant describes tsunamis as 'the underrated hazard'. Commenting on its effects on buildings, he notes that 'wood buildings offer no refuge from tsunami ... stone, brick or concrete block buildings will withstand flow depths of 1–2 m but are destroyed by greater flows'.[12]

The starting point for architects and planners in ascertaining the risk of a tsunami is to obtain an inundation map of the area of interest (Fig. 15.4). With an appreciation of the uncertainties and assumptions that affect the accuracy of such information, damage mitigation measures can be considered. The number of options appears to be limited to the construction of tsunami walls or barriers, planting dense areas of low trees, and relocation.[13] The Japanese have protected many fishing villages along the Sanriku Coast by installing massive reinforced concrete walls. One, over 12 m high with 1.5 m thick concrete walls, incorporates a traffic tunnel within its 8.5 m width. A very expensive option with considerable adverse environmental impacts, it is far more effective than wide plantings, which, although absorbing some of the tsunami energy, also add to the volume of water-borne debris. Relocation of tsunami-affected settlements has been undertaken in Petropavlovsk, Russia and in Hilo, Hawaii, which had been almost totally destroyed twice before much of its business district was rebuilt further inland.

Tsunami early-warning systems and identification and provision of evacuation routes are also effective methods to reduce loss of life. But in areas where tsunami flow can inundate low-lying coastal land many kilometres inland, and with warning periods measured only in minutes, there is nowhere safe to flee. For many such 'at-risk' people, what are termed 'tsunami vertical evacuation sites' are the only chance of survival.

The primary requirement of a tsunami shelter is to accommodate evacuees above the expected inundation level. As far as the structural design properties of a shelter are concerned, it must first be designed to resist seismic forces from ground shaking and then checked that it can withstand the considerable hydrodynamic pressures plus impact forces from water-borne debris. Buoyancy forces, as well as the scouring away of soil from around and under foundations, also need attention. Due to the possibility of nearby petroleum storage tank rupture, the structure must also be fire-resistant. A tsunami exerts large horizontal forces on any walls placed normal to its flow. Therefore, structural walls are to be avoided and non-structural walls designed to collapse at sufficiently low forces that will not endanger the overall performance of the structure.

Having ruled out shear wall structures as suited to tsunami resistance and considering all the requirements above, building designers are left with but one structural option – reinforced concrete moment frames in both orthogonal directions. Sacrificial interior and exterior walls at ground, and possibly first floor levels, allow tsunami surges to pass safely between columns. Allowance must be made for debris that might act as a temporary dam. One case study shows that a twelve-storey moment frame building can survive forces from a three metre tsunami flow depth without additional strength.[14] Lower buildings, with lesser design wind and seismic forces, are far more vulnerable. Guidelines currently under development for structural design criteria of multi-storey buildings capable of withstanding the forces applied by a tsunami have been given impetus in California by recent recommendations of its Seismic Safety Commission.[15]

FIRE FOLLOWING EARTHQUAKE

Another underrated and neglected seismic hazard, at least in the minds of architects designing buildings, are the fires that so often follow

▲ **15.5** A localized example of fire damage following the 1995 Kobe, Japan earthquake.
(Reproduced with permission from Adam Crewe).

a damaging earthquake. Charles Scawthorn and others warn: 'That large fires following earthquakes remain a problem is demonstrated by ignitions following recent earthquakes such as the 1994 Northridge and 1995 Kobe earthquakes (Fig. 15.5).[16] They recall that during the 1906 San Francisco earthquake fire, 28,000 buildings were lost over a period of three days and approximately 3000 people were killed. Yet that tragedy was far surpassed after the 1923 Tokyo earthquake where 77 per cent of the 575,000 buildings lost were destroyed by fire, and over 140,000 lives lost.

Post-earthquake fires are not inevitable. The widespread use of non-combustible building materials was probably the main reason that fires have been absent from several quake stricken areas including Bhuj, India, in 2001 and Marmara, Turkey in 1999, although one petroleum refinery there was destroyed by fire. The time of day and prevailing weather conditions also explain the absence of post-earthquake fires. If an earthquake strikes during windy weather fire spread is far more likely and rapid, as shown by past experience and computer fire modelling.[17]

How should planners respond to the threat of fire following earthquake? First, they should adopt best practice standards for reducing risk of fire spread from non-earthquake fires. Then, realizing how quickly post-earthquake fires can become conflagrations, they should recommend even higher standards. Wide streets are effective as firebreaks and less likely to be blocked by fallen rubble; urban parks provide places of refuge, and any reduction in the combustibility of building claddings and roofs will reduce fire risk. Existing buildings could even be required to comply with current fire-spread regulations. If the fire hazard of a community is assessed as very serious, measures as extreme as constructing purpose-designed buildings to function as fire barriers might be considered (Fig. 15.6).

▲ **15.6** Two of an irregular line of eighteen apartment blocks forming a firewall protect a refuge area. During a fire, metal shutters cooled by falling water protect the glazing. Shirahige, Tokyo.
(Reproduced with permission from Geoff Thomas).

Architects should advise their clients of the risk posed by post-earthquake fire. In cases of very high risk a client might request additional fire protection, such as enhanced fire-resistant cladding and shutters to cover exposed windows. Clients should also be advised to provide on-site water storage for fire sprinklers and provide hand-held fire appliances. During the 1995 Kobe, Japan earthquake 40 per cent of the post-earthquake fires were extinguished by the public.[18] The risk of gas ignitions can be reduced by providing flexible connections between buried pipes and buildings to accommodate relative movement without rupture. Otherwise, one can do little else other than meeting the various mandatory requirements for passive and active fire control systems, and voluntarily supporting efforts to achieve and maintain an effective fire-fighting force.

INTERDISCIPLINARY INTERACTION

When urban planners respond to the general seismic hazard posed by ground shaking, as well as to the specific earthquake-induced hazards mentioned above, they must work as members of interdisciplinary teams. Due to the complexity of seismic damage scenarios, planners need experts from other disciplines to help them refine their proposals and to check for any unintended negative effects that might arise.

Teresa Guevara-Perez warns how seismic vulnerability can increase when planners neglect to consult other disciplinary experts. She cites a case where city planners introduced regulations to increase ground floor parking but unintentionally promoted serious soft-storey configuration defects in the building stock.[19] Setbacks, as defined in Chapter 9, are another example. Some urban planning regulations encourage their use, even though they lead to irregular configurations with inferior seismic performance. An especially serious case of adverse seismic consequences arising from planning regulations is evident in Turkey. Before the introduction of the new Turkish Earthquake Code in 1998 it was common practice for first level floors and above to cantilever 1.5 m beyond ground floor columns. This practice, which leads to discontinuous perimeter columns, is considered one of the main reasons for the seriously damaged multi-storey buildings in recent Turkish earthquakes.[20]

REFERENCES AND NOTES

1 Coburn, A. and Spence, R. (2002). *Earthquake Protection (2nd Edn)*. John Wiley & Sons.

2 Californian Geological Survey (2003). The State of California Seismic Hazard Zones Map, *California Department of Conservation*, California.

3 Most earthquake damaged cities are rebuilt on the same sites. One exception is the relocation of the Scicilian town of Noto some 7 Km away, as

described in Tobriner, S. (1982). *The Genesis of Noto*. University of California Press.

4 Bahrainy, H. (1998). Urban planning and design in a seismic-prone region (the case of Rasht in Northern Iran). *Journal of Urban Planning and Development,* ASCE Urban Planning and Development Division, **124**:4, 148–180.

5 Mehdian, F., Naderzadeh, A. and Moinfar, A.A. (2004). A comprehensive master plan study on urban seismic disaster prevention and management for Tehran City. *Proceedings of the 13th World Conference on Earthquake Engineering,* Vancouver, August 1–6. Paper no. 913, 14 pp.

6 Balamir, M. (2004), Urban seismic risk management: the earthquake master plan of Istanbul (EMPI). *Proceedings of the 13th World Conference on Earthquake Engineering,* Vancouver, August 1–6. Paper no. 9005, 22 pp.

7 Kuroiwa, J. (2006). Peru's sustainable cities program 1998-2005 and its application to large built-up areas. *Proceedings of the 8th US Conference on Earthquake Engineering,* April 18–22, San Francisco. Paper no. 269, 10 pp.

8 Hosseini, M and Shemirani, L.N. (2003). The role of urban planning and design in lifeline-related seismic risk mitigation. In Advancing mitigation technologies and disaster response for lifeline systems. *Proceedings of the Sixth US Conference and Workshop on Lifeline Earthquake Engineering.* August 10-13, Long Beach, California, Technical Council on Lifeline Earthquake Engineering, Monograph No. 25, American Society of Civil Engineers, 779–788.

9 Lew, M. and Real, C.R. (2004). Implementation of the seismic hazards mapping act in California. *Proceedings of the 13th World Conference on Earthquake Engineering,* Vancouver, August 1–6. Paper no. 3166, 15 pp.

10 Van Dissen, R. *et al.* (2006). Mitigating active fault surface rupture hazard in New Zealand: development of national guidelines, and assessment of their implementation. *Proceedings of the 8th US Conference on Earthquake Engineering,* April 18–22, San Francisco. Paper no. 633, 10 pp.

11 Lewis, L. (2007). Inside the nuclear plant hit by earthquake. *The Times,* August 9.

12 Bryant, E. (2001). *Tsunami: The underrated hazard*. Cambridge University Press.

13 Dudley, W.C. and Lee, M. (1998). *Tsunami* (2nd Edn). University of Hawaii Press.

14 Yeh, H., Roberston, I. and Preuss, J. (2005). *Development of design guidelines for structures that serve as tsunami vertical evacuation sites.* Washington State, Department of Natural Resources.

15 State of California eismic Safety Commission (2005). *The Tsunami Threat to California: Findings and recommendations on tsunami hazards and risks,* CSSC 05-03. State of California Seismic Safety Commission.

16 Scawthorn, C., Eidinger, J.M. and Schiff, A.J. (2005). *Fire Following Earthquake.* Technical Council on Lifeline Earthquake Engineering monograph; no. 26, American Society of Civil Engineers.

17 Thomas, G.C. *et al.* (2003). In Evans, D. D (ed.), Post-earthquake fire spread between buildings: estimating and costing extent in Wellington). *Fire safety*

science–proceedings of the seventh international symposium, International Association of Fire Safety Science, 691–702.

18 Thomas, G.C. (2005). Fire-fighting and rescue operations after earthquakes–lessons from Japan. *Proceedings of the New Zealand Society for Earthquake Engineering Annual Conference,* Wairakei. CD-ROM, 8 pp.

19 Guevara-Perez, L.T. (2006). Urban and building configurations of contemporary cities in seismic zones. *Proceedings of the 8th US Conference on Earthquake Engineering,* April 18–22, San Francisco. Paper no. 1775, 10 pp.

20 Sesigür, H. *et al.* (2001). Effect of structural irregularities and short columns on the seismic response of buildings during the last Turkey earthquakes. In *Earthquake Resistant Engineering Structures III, International Conference on Earthquake Resistant Engineering Structures* (3rd) WIT, Southampton, Corz, A. and Brebbia, C.A.

16 ISSUES IN DEVELOPING COUNTRIES

INTRODUCTION

Designers who practice in seismically active developing countries face numerous unique issues with respect to earthquake-resistant design and construction. Situations where widespread poverty coexists with aspirations to advance technologically and economically lead to many extremes. Adjacent to a building project employing state-of-the-art seismic resistance, such as seismic isolation, basic engineered or non-engineered construction in all likelihood lacks any intentional earthquake resisting features. Although all of the content of previous chapters is applicable to developing countries, this chapter bridges between technologies and practices commonplace in so-called developed countries and what is rarely encountered and possibly resisted in developing countries.

Developing countries are characterized by rapid urbanization, most of which is uncontrolled or poorly controlled. Weak or non-existent regulatory environments, a subsequent lack of enforcement of design and construction standards, and a lack of effective technical and professional leadership, have led to seismically vulnerable building stocks. Polat Gülkan's description of the situation in Turkey applies elsewhere: '[the] quality of the country's building stock is highly variable and control and supervision of design and construction seem to have been pre-empted in the interest of a fast-paced rural to urban conversion'.[1] Sudhir Jain's insight into India's construction industry prompts him to assert that 'a huge number of unsafe buildings continue to be built every day in different cities and towns'. By way of illustration, he notes how approximately three-quarters of 6000 pre-cast concrete school buildings constructed between 1999 and 2000 collapsed or were seriously damaged during the 2001 Bhuj earthquake (Fig. 16.1).[2]

▲ **16.1** A damaged precast concrete school building due to failure of roof panel connections. 2001 Bhuj, India earthquake. (Reproduced with permission from IIT Bombay Team).

▲ **16.3** Frame and panel buildings in Tashkent that may not meet current seismic standards.
(Reproduced with permission from Nadira Mavlyanova).

▲ **16.2** Seismically vulnerable buildings in Mumbai.
(Reproduced with permission from R. Sinha).

Developing country researchers who assess the seismic vulnerability of their own cities come to similarly troubling conclusions. After noting that almost 50 per cent of Mumbai's population inhabits informal slum houses, Sinha and Adarsh conclude that 'the occurrence of a code-level (MSK Intensity VII) at Mumbai may lead to massive loss of life and damage of buildings. Depending on the time of day, between 25,000 and 42,000 people may perish due to structural collapse and damage in the earthquake. The numbers of serious injuries may also range between 71,000 to 118,000, possibly placing a very severe strain on the emergency relief and health-care infrastructure. Similarly, a very large number of buildings (in millions) may be damaged or lost'[3] (Fig. 16.2). Although not as dramatic in its findings, a report on the vulnerability of Tashkent, the capital of Uzbekistan states that, 'About 43 per cent of the inhabitants of the Tashkent city live in buildings that were not adequately designed and constructed to meet current standards of seismic resistance'.[4] In this case, the most hazardous buildings are not so-called 'informal' buildings, but engineered nine to sixteen-storey frame-panel buildings (Fig. 16.3). They were prefabricated and connected by field welds of poor quality.

It comes as no surprise that informal or non-engineered buildings are seismically deficient. Built from heavy, brittle and often weak materials they lack any tension elements to tie walls together, or walls to floors or roofs, and that might strengthen walls against in-plane shear failure. Unfortunately, this seismically vulnerable construction, which may be new, has in many cases replaced traditional building types which have demonstrated better seismic performance during past quakes. However, it *is* unexpected that engineered buildings, which may also be designed by architects, are so vulnerable. The reality is that the seismic performance of many engineered buildings in developing countries is compromised by configuration, design and construction defects. In most countries a sound earthquake-resistant building can be achieved only by going well beyond their conventional practice, to the extent of introducing quite different design and building techniques.

DESIGN

The extremes found in all sectors of developing countries are manifest in the quality of the engineering design and the construction of buildings. Particularly in major cities, some engineering consultants practice high levels of seismic expertise. But overall, the standard is low and in most projects seismic engineering input is non-existent. And as noted previously, engineering input is no guarantee of satisfactory seismic performance. Apart from a structural engineer's personal competence in seismic design, which may be dubious given a lack of seismic design content in schools of engineering curricula, other issues that can reduce the seismic resilience of buildings include code requirements, configuration irregularities, design detailing and quality control of construction.

In many developing countries, code design force levels are low by international standards. After investigating the 2003 Boumerdes, Algeria earthquake, Fouad Bendimerad comments, 'While the Algerian earthquake code prescribed design values for buildings in the order of 15 per cent gravity, there is evidence that in the epicentral region both the horizontal and vertical accelerations from the earthquake exceeded 100 per cent of gravity'.[5] That some affected interest groups in the building industry, such as building developers and others concerned about building affordability, resist seismic design forces being increased beyond those specified in existing out-dated codes, is understandable. But the consequences of low design forces are weak buildings. To survive a damaging quake, such buildings require an unrealistically high level of ductility. Designers should be aware of the return period of the design-level earthquake of their own code. Only then can they advise their clients of the seismic risk to a building during its

▲ **16.4** Soft storey building, Venezuela.
(Reproduced with permission from Wiss, Janney, Elstner Associates, Inc.).

anticipated design life. As discussed in Chapter 13, some clients are willing to pay extra for improved seismic protection.

A review of recent earthquake damage to buildings in developing countries has found that the majority of seismic vulnerability of engineered buildings arise from configuration irregularities.[6] This finding aligns with that of a Venezuelan study which, while acknowledging the existence of a modern seismic code, concludes that 'significant conceptual errors in the design of the lateral force-resisting systems of new buildings are recurring on a near-universal level, often as a result of ignoring the potential adverse effects of nonstructural elements on the structural system'.[7] The study's list of primary deficiencies include those discussed in Chapters 8 and 9; namely, soft and weak storeys, short columns, strong beam–weak columns and torsion (Fig. 16.4).

Masonry infill walls that are not separated from moment frames are usually the main cause of each of these configuration problems. The only way to overcome them is by introducing the techniques of Chapter 10 which, for most countries, represent new building practices. The two most practical solutions for developing countries to improve the seismic performance of their new buildings are to use reinforced concrete shear walls to resist seismic forces and to adopt confined masonry construction. Unfortunately, both solutions reduce ground floor openness and transparency and inevitably entail greater construction cost when designed and built properly. Several leading structural engineers with developing country experience are of the view that reinforced concrete shear walls possess significant advantages over moment frames from a seismic perspective.[5,8] The historic seismic performance record of shear walls is far better and, due to their less sophisticated design, detailing and construction, they are more dependable. Alternatively, a confined masonry structural system, as outlined in Chapter 5, can be adopted. But its structural capabilities restrict it to low- to medium-rise construction only and necessitate rather rigorous limitations on layout, openings and structural footprint.

The provision of more rational seismic structural systems is another area where developing countries need to consider a new approach to how they build. More regular column orientation in-plan (see Fig. 5.34) and realistic column dimensions are required. Moment frames

▲ **16.5** Slender columns can not provide adequate seismic resistance. Mumbai.
(Reproduced with permission from R. Sinha).

▲ **16.6** Inadequate welds caused three brace failures. Bam, 2003 Bam, Iran earthquake.
(Reproduced with permission from Jitendra Bothara).

incorporating 'stick' columns, commonly around 230 mm square are too weak and flexible to function as seismic resisting elements. Such small cross-sectioned members should have no more expected of them than to resist gravity forces (Fig. 16.5). As illustrated in Fig. 5.44, moment frames need to be designed using the Capacity Design approach that results in columns stronger than beams – the so called 'weak beam–strong column' approach.

Designers need to be prepared for initial negative reactions to the introduction of sound earthquake-resistant practices. Clients who are used to columns fitting within the thickness of partition walls may be mildly shocked at the greater sizes of Capacity Designed columns required to ensure adequate seismic performance. Changes to traditional approaches, which are strongly embedded in a country's construction culture, require considerable justification and professional insistence.

An architect needs to choose a structural engineer for his or her project carefully. Both need to present a united front when challenging and changing traditional ways of building. The engineer must be amenable to discarding traditions that have proven seismically inadequate and to adopt new approaches. He or she should be able to demonstrate their personal technical competence, not only in conceptual seismic design, but also by how they deal with detailed design. Unlike many other structural design situations, structural performance under seismic forces is very sensitive to the quality of detailing.[9] In reinforced concrete design, reinforcing detailing is extremely important, just as welding details are crucial for steel construction (Fig. 16.6). As illustrated

▲ **16.7** Incorrectly bent column ties led to this column almost collapsing. 1987 Tarutung, Indonesia earthquake.

in Figs 12.5 and 16.7, such an apparently small detail as incorrectly bent column ties can lead to the collapse of an entire building. Too often, this and other reinforcing factors are detailed by draughtsmen not trained in seismic detailing and who do not appreciate the critical importance of their work.

CONSTRUCTION

Maintaining adequate standards of construction is a challenge in any country. But a relatively uneducated workforce, and a less than robust construction quality control regime, common in developing countries, leads to low standards. A case involving poor site welding has already been mentioned, and earthquake reconnaissance reports frequently note how low concrete strengths and deficient reinforcement detailing contribute to the collapse of reinforced concrete buildings. In engineered construction material property specifications need to be upheld. An architect needs to be aware of these and other potential construction problems. Also as a general principle, seismic resistant construction should be kept as simple as possible.

RESOURCES

Two organizations lead the way in providing seismic design information relevant to building designers in developing countries. The main resource of the World Housing Encyclopaedia is its collection of more than 100 reports of housing types from 37 different countries.[10] The reports, which include brief notes on the cultural settings of buildings and their inhabitants by describing many non-engineering aspects, focus their attention upon the earthquake resistance of buildings. In most cases expert commentaries outline the seismic weaknesses of construction. Solutions and guidelines for improving the seismic resistance of various construction types are found in tutorials, manuals and other on-line documents.

The National Information Center of Earthquake Engineering, based at the Indian Institute of Technology, Kanpur, India, is another prolific provider of information relevant to engineers and architects in developing countries.[11] A wide range of publications suitable for both practitioners and educators includes international recommendations for improving the seismic resistance of non-engineered construction.[12] The Earthquake Hazard Centre, Victoria University of Wellington, also disseminates earthquake damage mitigation information. It publishes a quarterly newsletter with information and articles chosen for their relevance to developing countries.[13] Finally, mention should be made of Julio Kuroiwa's contribution.[14] His chapter on earthquake engineering, which draws mainly upon earthquake damage and earthquake resistant developments in Latin America, is notable for its breadth of coverage and relevance to developing countries.

REFERENCES

1 Gülkan, P. (2000). Code enforcement at municipal level in Turkey: failure of public policy for effective building hazard mitigation? *Proceedings: Managing earthquake risk in the 21st century*, Nov. 12–15, Palm Springs, California, U.S.A., 895–900.

2 Jain, S.K. (2005). The Indian earthquake problem. *Current Science*, **899**, 1–3. Sinha, R. and Adarsh, N. (1999). A postulated earthquake damage scenario for Mumbai. *ISET Journal of Earthquake Technology*, **36**:2–4, 169–183.

4 Mavlyanova, N. et al. (2004). Seismic code of Uzbekistan. *Proceedings 13th World Conference on Earthquake Engineering*, Vancouver, B.C., Canada, Paper no. 1611, 10 pp.

5 Bendimerad, F. (2004). The 21 May 2003 Bourmerdes earthquake: lessons learned and recommendations. *Proceedings 13th World Conference on Earthquake Engineering*, Vancouver, BC, Canada, Paper no. 9001, 10 pp.

6 Charleson, A.W. and Fyfe, G.D. (2001). Earthquake building damage in developing countries: a review of recent reconnaissance reports. *Bulletin of the New Zealand National Society for Earthquake Engineering*, **34**:2, 158–163.

7 Searer, G.R. and Fierro, E.A. (2004). Criticism of current seismic design and construction practice in Venezuela: a bleak perspective. *Earthquake Spectra*, **20**:4, 1265–1278.

8 Sauter, F.F. (1996). Design philosophy for seismic upgrading of buildings: a Latin American perspective. *Proceedings of the 11th World Conference on Earthquake Engineering*, Acapulco, Mexico, Paper 2098, 10 pp.

9 Charleson, A.W. (1998). The importance of structural detailing in seismic resistant construction. *Proceedings of the Eleventh Symposium on Earthquake Engineering*, Roorkee, India, 373–379.

10 World Housing Encyclopaedia. Earthquake Engineering Research Institute, Oakland, California, http://www.world-housing.net/index.asp

11 National Information Centre of Earthquake Engineering (NICEE), Indian Institute of Technology, Kanpur, India, www.nicee.org

12 International Association of Earthquake Engineering (IAEE). (1986). *Guidelines for earthquake resistant non-engineered construction*. The International Association for Earthquake Engineering, Tokyo, Japan

13 Earthquake Hazard Centre, Victoria University of Wellington, New Zealand, http://www.vuw.ac.nz/architecture/research/ehc/

14 Kuroiwa, J. (2004). *Disaster reduction: living in harmony with nature*. Editorial NSG S.A.C., Lima.

17 EARTHQUAKE ARCHITECTURE

INTRODUCTION

'Earthquake architecture' describes the architectural expression of some aspect of earthquake action or resistance. The breadth of expressive possibilities inherent in earthquake architecture can be considered conceptually as two paths. In the first path, often characterized by a straightforward exposure of necessary seismic technology, seismic resisting structure is integrated and expressed architecturally. Structural form and details may not only celebrate the roles they play in resisting seismic forces but also provide well-configured seismic resisting systems. The second path of earthquake architecture involves journeying with architectural concepts containing metaphoric and symbolic references to seismic issues. Examples from this path typically include buildings whose architectural forms have been distorted in response to, for example, earthquake destruction or geological forces, and whose seismic performance may not be as predictable as regularly configured buildings. These two paths are considered separately in more detail in the following sections.

Why then might one engage in earthquake architecture? Some architects taking the first path wish to openly acknowledge the necessity to safeguard a building against seismic damage. They may opt for a show of strength, exposing a muscular structure not designed to outwit a quake by stealth but by strength. There may be a desire to explore the potential for aesthetic richness through a celebration of seismic technology. This appeared to be the motivation of architect David Farquharson who introduced innovative seismic resistant features in South Hall, University of California at Berkeley, in 1873. He believed that safety features of a building should be revealed to passers-by in the form of art, and proposed a method that integrated reinforcement with decorated wrought iron work.[1] For other architects, rather than basing their design concepts on international architectural trends unrelated to the cultural and other settings of their own countries, they generate a regional architectural response given the geophysical setting of their region and site via metaphorical and symbolic references.

While there are some interesting and attractive examples of earthquake architecture internationally, we need to remember that earthquake architecture, unlike seismic resistant design, is optional.[2] Since earthquake architecture has the potential to aesthetically enrich architectural form it definitely warrants exploration and development. Yet the primary seismic focus of architects must always be upon ensuring buildings possess adequate seismic resilience. Tadao Ando emphasizes this point during his 1997 Royal Gold Medal Address: 'Returning to Kobe and visiting the site of the earthquake, the first thing that struck me was just how important a responsibility we architects have on the very basic level of providing safety and security for people'.[3]

Earthquake architecture largely describes what *might be* rather than what *is*, given a general lack of architectural expression of seismic design in earthquake-prone regions. Christopher Arnold notes that earthquake architecture is not an established architectural movement. Probably with the second path of earthquake architecture in mind he suggests that the reason 'may be due to the psychological desire to deny the prevalence of earthquakes: building designs which remind the knowledgeable observer are striking a negative note'.[4]

Most buildings do not openly reveal the fact that they have been designed to resist seismic forces. Occasionally, seismic (and wind) systems are expressed on building façades such as by external braced frames, but often the architectural aim of structural exposure is to emphasis verticality. Columns may be the only visible structural members. Sometimes beams are hinted at by opaque cladding panels but frequently, as in the case of reflective glass façades, all structure is concealed. Separation gaps between buildings and precast panels required to accommodate earthquake-induced drifts may be visible, but no attempt is made to celebrate aesthetically these necessities. It is probably in the context of seismically retrofitted buildings where earthquake resistant technology is most apparent. However, as discussed in Chapter 12, architects and even some engineers are concerned by the crudeness of much of this work. Highly visible diagonal steel members inserted seemingly without regard to a building's existing architecture, image and function, do little to commend an earthquake architecture approach.

The chapter continues by outlining the wide range of areas within architectural design practice where earthquake architecture might be realized. After exploring the potential for expressing structural principles and actions as earthquake architecture, several case-studies are discussed briefly. Then the limitless extent to which seismic issues can generate design concepts involving seismic metaphor and symbolism are noted.

Expression of seismic resistance

Opportunities for expressing seismic resistance present themselves in most aspects of architecture, from an urban scale through to detailed design. This section considers how seismic resistant measures might be visually expressed in different architectural settings and elements.

Urban planning

As discussed in Chapter 15, seismicity should impact urban planning. Examples include: sites which are underlain by active faults subject to surface rupture – these should remain undeveloped; open spaces provided for safe refuge from damaged buildings and the temporary storage of post-earthquake debris; and wide streets provided to ensure access for emergency response personnel. Landscape architecture elements can also enhance public safety. Despite strong client pressure, Frank Lloyd Wright provided a feature pool in front of the Imperial Hotel, Tokyo. He argued that it would be a reservoir from which to fight post-earthquake fires and describes it being put to exactly that use in the immediate aftermath of the 1923 Tokyo earthquake.[5]

At a more detailed level of urban design, seismicity is acknowledged by building back from site boundaries, creating vertical separation gaps between buildings to avoid pounding. These gaps are usually hidden. But might they be expressed and even have their important safety function celebrated?

Building form and massing

Possibilities exist for building form to respond to earthquake architecture ideas. On one hand, pyramidal and squat building forms convey a visual sense of stability in the face of seismic forces. On the other hand, very tall buildings might express how, to a significant degree, they self-isolate themselves from earthquake shaking, due to their long natural periods of vibration. At a finer scale, it is possible to alleviate potential building configuration problems, such as set-backs, or soft-storeys, with structural members that are significant architectural elements themselves. Consider, for example, a building with huge one-bay double-height moment frames to avoid a soft-storey (see Fig. 6.8).

As noted above, façades offer expressive opportunities. The degree of visual exposure of structural members can range from the subtle indication of structure to its direct expression. Conventional braced frames, a prevalent exposed seismic system at least in retrofitted buildings, are probably overworked but well-designed variants can make a positive contribution to a cityscape. The two reinforced concrete braced cores at each end of the building shown in Fig. 5.29 are an example of refined concrete braced frames above a solid potential plastic hinge region.

Moment frames are also exposed frequently but often their seismic resisting function is not obvious. The relatively massive dark coloured wind and seismic frame in Fig. 6.8 contrasts with slender and white gravity frame structures on either side.

▲ **17.1** Seismic braces define the main circulation route of this retrofitted building. Educational building, San Francisco.

Interiors

An exposed interior structure can contribute to spatial quality and aesthetic richness. Gravity force resisting elements such as columns and beams feature most commonly, but there is every reason for interior seismic structure to be equally successful in contributing architecturally as well as structurally (Fig. 17.1).

At a detailed level, floor, wall and ceiling seismic separation gaps discussed in Chapter 8 also provide expressive architectural opportunities. An exposed sliding or roller joint between the base of stairs and floors is another example of the aesthetic exploitation of necessary seismic detailing (see Fig. 10.22).

Non-structural elements

Heavy and strong elements, such as precast concrete and masonry wall panels, may require special seismic separation; especially if the seismic resistance of their building is provided by relatively flexible moment frames. Opportunities exist to express the separation gaps between structure and cladding panels and between cladding panels themselves. Where cladding panel separation concepts are similar in principle to the action of flexible and scaly reptilian skins, details that express that action are worthy of investigation. Similarly, the provision for movement within seismic mullions may warrant design exploration (see Fig. 11.15). Even partition wall-to-structure connections designed to accommodate interstorey drift provide opportunities for appropriate aesthetic expression.

Seismic restraint of building contents is another area ripe for architectural expression. With exemplary design and detailing, attractive and elegant seismic restraints for items such as electronic appliances, bookshelves and office equipment might even enhance an interior environment.

Seismic hardware

There is increasing use of seismic isolators, structural fuses, dampers and bearings in modern buildings. Figure 17.2 illustrates mild steel cantilever dampers that are exposed around the perimeter of a building. This is possibly the first time that seismic isolation has been articulated to any significant degree. Such a design approach is worthy of further development. In most seismically isolated buildings, exciting and innovative technologies are hidden from public view and their expressive qualities, like those of devices such as lead-extrusion dampers, are wasted (see Fig. 14.18). Architects in seismic regions also have yet to connect their commonly expressed wish to 'float a building' with the aesthetic opportunities offered by seismic isolation.

▲ **17.2** A steel damper in the form of a horizontal tapered cantilever on a building perimeter. Office building, Auckland.

EXPRESSION OF STRUCTURAL PRINCIPLES AND ACTIONS

This section explores the potential for expressing seismic principles architecturally in the context of common structural systems. Depending on the depth of seismic knowledge within a design team of architects and engineers, layers of increased sophistication in understanding can be accessed in order to realise numerous manifestations of earthquake architecture. Typically, a high degree of structural and

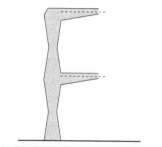

(a) Part-elevation of seismic moment frame

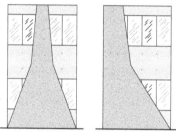

(b) Elevations of shear walls

▲ **17.3** Haunched members of a moment frame (a) and tapered shear walls (b) reflect their seismic bending moment diagrams.

▲ **17.4** High strength fibre composite material wraps and confines potential column plastic hinge areas. Wellington.

architectural collaboration and integration is necessary to fully exploit these possibilities for the purpose of architectural enrichment.[6]

Moment frames

At the most basic level, structural actions such as the axial forces, bending moments and shear forces arising from seismic forces can be expressed in the detailing or shaping of structural members. Usually the choice of just one action, such as bending, provides enough architectural potential (Fig. 17.3). If this were the chosen action, then one consequence for beam and column detailing is that members are haunched or tapered from beam-column junctions to achieve minimum cross-sectional depths at their mid-spans (Fig. 17.3(a)). Shear walls shaped to reflect their seismic actions also offer interesting possibilities (Fig. 17.3(b)).

A further level of detailing sophistication can express the concept of Capacity Design (Chapter 3). The fundamental requirement for moment frames is for seismic damage to columns to be suppressed and damage confined only to beams. The concept of strong columns–weak beams can be expressed easily. Architectural attention can also be paid to the expression of potential plastic hinges at ends of beams. Engineers anticipate and design for concentrated damage in these areas. In reinforced concrete construction, beam cores at the ends of beams are confined by closely-spaced ties wrapped around the horizontal reinforcing. Provision of this confinement conjures up images of binding, strapping or bandaging. The retrofitting of some existing earthquake-prone structures involves wrapping column plastic hinge zones with high-strength materials (Fig. 17.4). The utilitarian nature of the solution illustrated here does not preclude more elegant alternatives for new construction.

For glue-laminated wood frames architects can use ductile steel beam-column joints to articulate the structural goal of preventing damage to the brittle wooden members (see Fig. 5.40).

There are other opportunities to integrate specific reinforcement detailing requirements with architectural expression. Take the beam stubs projecting from corner columns that were popular in New Zealand multi-storey frame buildings in the 1970s. Stubs solved the problem of adequately anchoring top and bottom beam longitudinal bars, and lessened reinforcement congestion in the beam-column core,

▲ **17.5** Typical beam studs. Educational building, Wellington.

easing concrete placement (Fig. 17.5). Recent research has led to other satisfactory anchorage methods.

Braced frames

Apart from structural fuse regions between the inclined braces of eccentrically braced frames, braced frame members primarily resist forces by tension or compression. Variations in axial force levels up a building can be expressed, as in Fig. 5.29. Towards the top of the building, less shear force and bending moment permit member cross-sections to be reduced.

Capacity Design considerations once again provide more profound detailing opportunities. An economical and reliable approach is for fuse areas in tension and/or compression members, that are selected to yield, to be deliberately weaker than surrounding members. Other members and connections will therefore remain undamaged during an earthquake. There is considerable scope for structural fuses to be articulated architecturally, particularly if the detailing is refined and positioned where it can be appreciated (see Fig. 5.31).

Shear walls

Bending moments and shear forces are the dominant seismic actions within walls. Possible expressive architectural strategies include differentiating between wall chords and webs; that is, between moments and shear forces (Fig. 17.6). Chord dimensions and wall thicknesses can be

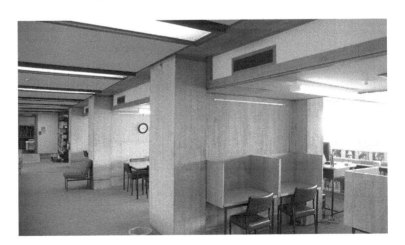

▲ **17.6** Reinforced concrete shear walls where the wall chords are differentiated from the webs. Educational building, Wellington.

▲ **17.7** A steel cross-braced diaphragm at first floor level. Library, Palmerston North, New Zealand.

varied in response to force intensity. One approach that articulates varying intensities of shear force and the need for a plastic hinge at a wall base without penetrations, has been to increase the area of fenestration towards the top of a wall (Fig. 5.8). But far more explicit examples of shear walls expressing their structural actions are possible (Fig. 17.3(b)).

Horizontal elements

Diaphragms, together with other members that collect and transfer forces into vertical structure, are essential components of seismic resisting systems. As discussed in Chapter 4, diaphragms resist shear forces and bending moments. Collector and tie members act in tension and compression. They may appear as beams or ribs or can be articulated separately. Usually floor slabs function as diaphragms, but across large openings special diaphragm structure, such as steel or concrete cross-bracing, can be expressed (Fig. 17.7).

▲ **17.8** Sketch of a segment of the façade of the Nunotani Headquarters building, Tokyo.

SEISMIC ISSUES GENERATING ARCHITECTURE

Travellers on the second path of earthquake architecture explore numerous ways that metaphor and symbolism can inform architectural responses to seismic issues. Christopher Arnold cites the example of the Peter Eisenman designed Nunotani Headquarters Building in Tokyo, completed in 1992 (Fig. 17.8). Disjointed and displaced façade elements were intended to 'represent a metaphor for the waves of movement as earthquakes periodically compress and expand the plate structure of the region'.[4] Once aware of the design idea, one can perceive seismic activity in main elevations of the building; but some viewers might interpret the architectural distortions as either evidence of seismic damage itself or even incompetent construction. It is not known to what extent the unusual form of the building that responded admirably to the client's brief for an 'aggressive, contemporary image', contributed to its demolition only eight years later.[7]

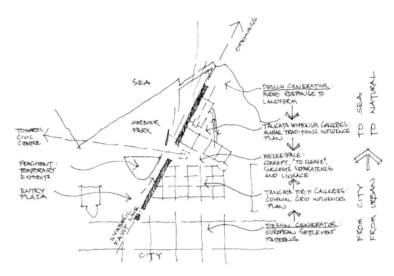

▲ **17.9** Concept plan of the Museum of New Zealand Te Papa Tongarewa, Wellington. (Reproduced with permission from Pete Bossley).

A lesser known example of seismic issues informing architectural design occurs in the Museum of New Zealand Te Papa Tongarewa, Wellington. The design architect explains:

> The need for direct connections ... in turn led to the introduction of the idea of geological power/Ruaumoko [the Maori god of earthquakes and volcanoes] expressed as a mighty Wall slicing diagonally through the building. This symbolic fault line (parallel to the actual earthquake fault line nearby, on the western side of the harbour) created a fissure of space which houses the newly created Entry from the city (Figs 17.9 and 17.10).[8]

More so than the first example in this section, it is unlikely that visitors to the building mentally link the primary architectural element; in this case, a highly penetrated wall, with the underlying concept – a fault line. Perhaps if the concept had been developed further by incorporating other references to fault movements, such as vertical displacement and a non-vertical fault-plane orientation, less interpretation would be needed. However, the fact remains that seismic issues have provided the inspiration for an innovative architectural design concept.

Countless other metaphors are possible sources of design ideas for architects who desire to recognize, to some degree at least, tectonic activity adjacent to their site and who are grappling with the development of building form. Apart from ideas of crustal compression and expansion, a fuller list includes slicing, fragmentation (also acknowledged

▲ **17.10** An exterior view of the Museum of New Zealand showing the penetrated 'wall' that symbolizes an adjacent fault line, Wellington.

in the Te Papa design), splitting, fracturing, sliding, folding and faulting. Geological metaphors have also been adopted by architects in non-seismic regions and developed into central design concepts.

A design studio program at the Victoria University of Wellington School of Architecture provides another example of how seismic issues generate innovative architectural form. In 2002 a group of eight fourth-year students participated in a studio program centred upon 'how notions of earthquake architecture could inform the design process'.[9] After researching earthquake effects and actions, and considering them as architectural propositions, various design ideas were identified, some of which are listed below.

Geology and seismology

- Seismic waves
- Faulting
- Earthquake-affected landforms
- Contrast between geological and seismograph time-scales

Construction issues

- Post-earthquake propping
- Tying elements together
- Post-earthquake ruins
- Seismic resisting technology and componentry
- Contrast between gravity and seismic resisting structure

Other earthquake-related ideas

- Temporary buildings for disaster relief
- Seismograph
- Expression of structural actions
- Brittle behaviour
- Plastic behaviour

Ideas not specifically related to earthquakes

- Healing processes like scabs that form after an injury
- External forces acting on a building
- Insecurity
- Preparedness
- Engineer-architect relationship

After students tested one of their ideas in the context of a small public building – a suburban library – they designed a more complex building.

▲ **17.11** The building form expresses earthquake damage and subsequent repair/propping.
(Reproduced with permission from Luke Allen).

▲ **17.12** One section of the final design model.
(Reproduced with permission from Lebbeus Woods).

Adopting another of their ideas, they had to develop its architectural potential. The brief required a medium-rise building on an urban corner site to accommodate purpose-designed offices for an innovative earthquake engineering consultancy. Figure 17.11 illustrates one of the student schemes.

At a more theoretical level, Lebbeus Woods draws upon earthquake destruction in his investigations of architectural transformation.[10] He begins with abstract sketches reminiscent of a featureless landscape subject to extreme lateral-spreading or littered by post-tsunami debris. Shard-like forms that convey fracturing, fragmentation and catastrophic movement are then expressed by line drawings. A dense pattern of straight lines devoid of any orthogonality loses none of its expression of the flow of debris. These patterns are then physically modelled. They read as a dense and chaotic cityscape consisting of thin slithers of buildings separated by similarly shaped interstices (Fig. 17.12). Alluding to damage caused by the 1994 Northridge and the 1995 Kobe earthquakes Woods explains the background and aims of the project:

'In the light of the consistent failure of leading societies such as the United States and Japan to build effectively against earthquake, it is reasonable to reconsider the dominant philosophies, techniques and goals of building and urban design in earthquake regions. As at this writing, such a reconsideration by architects and planners has hardly begun. Few efforts go beyond the defensive "reinforcement" of existing conceptual and physical structures, or have the ambition to open up genuinely new possibilities for architecture in relation to the earth's continuing process of transformation'.[10]

While this work is certainly an exciting visual exploration of how earthquake destruction can be transformed into architectural form, its outcome is unsatisfactory from a seismic safety perspective. Narrow streets and slender shard-like buildings are the antithesis of sound seismic design principles. The value of the project lies in its challenge to others to undertake their own 'reconsiderations'. Other creative methods and attempts need to explore the possibilities for a radical architecture that not only expresses seismicity in an aesthetic sense but is inherently more seismically resilient.

Arata Isozaki's Japanese Pavilion at the Venice Biennale: sixth International Architecture Exhibition, 1996, provides another example of seismic issues informing non-built architecture. Responding to the theme of 'The Architect as Seismograph' Isozaki devotes the entire pavilion to what was then the very recent devastating Kobe earthquake. In his response titled 'Fractures', noted by one reviewer as 'an act of caustic irony', mounds of post-earthquake debris are supplemented by photographs of damaged buildings. Commenting upon the exhibition, Isozaki states:

> '... I feel this focus on the ravages of the hard-hit city, rather than on some optimistic architectural proposal, to be a more accurate expression of the state of Japanese architecture today'.[11]

Luis Fernández-Galiano takes a very different stance. He sees instability of architectural form, a possible earthquake architecture concept, as a potential form of therapy. Accordingly, he perceives buildings like, for example, the CCTV Headquarters, Beijing (Fig. 6.25) that appear to be unstable, as sending out a reassuring message:

> '... if we can make these impossible forms stand, so will we manage to keep a fragile world stable. .... like a vaccine that injects debilitated pathogenic germs into the organism, unstable architecture provokes small commotion, controllable fractures and tamed calamities that feign danger through fatal forms and cauterize anxiety through cautious catharsis'.[12]

Architects are able to explore a large range of seismic-related issues in order to develop and enrich their architecture. Design possibilities in many different areas and scales of architectural practice abound.

REFERENCES AND NOTES

1 Tobriner, S. (1998). How has architecture responded to earthquake challenges over time? Engineering and architectural responses to the San Francisco earthquakes of 1868 and 1906. *Proceedings of the 50th Annual Meeting of the Earthquake Engineering Research Institute*, Feb. 4–7, San Francisco, 9–12.

2 For some examples of earthquake architecture, refer to Garcia, B. (2000). *Earthquake Architecture: New construction techniques for earthquake disaster prevention*. Loft Publications.

3 Ando, T. (1997). Tadao Ando's Royal Gold Medal Address, *Concrete Quarterly*, Autumn, 2–7.

4 Arnold, C. (1996). Architectural aspects of seismic resistant design. *Proceedings of the Eleventh World Conference on Earthquake Engineering*, Elsevier Science Ltd. Paper no. 2003, 10 pp.

5 Wright, F.L. (1977). *Frank Lloyd Wright: an autobiography*. Horizon Press.

6 For a discussion on how structure, as an element of architecture, can generally enrich architecture, see Charleson, A. W. (2005). *Structure as architecture: a sourcebook for architects and structural engineers*. Elsevier.

7 GA (1993). Eisenman Architects Nunotani Headquarters Building. *GA Document*, No. 37, 82–9.

8 Bossley, P. (1998). *Te Papa – an architectural adventure*. Te Papa Press, Wellington.

9 Charleson, A.W. and Taylor, M. (2004). Earthquake architecture explorations. *Proceedings of the 13th World Conference on Earthquake Engineering*, Vancouver, B.C., Canada. Paper 596, 9 pp.

10 Woods, L. (2001). *Earthquake! A post-biblical view*. RIEAeuropa Concept Series, Springer.

11 Isozaki, A. (1997). On ruins – the 1995 earthquake in Kobe, Japan, *Lotus*, **Vol. 93**, June, 34–45.

12 Fernández-Galiano, L. (2000). Earthquake and therapy. *Lotus*, **Vol. 104**, 44–47.

18 SUMMARY

Although the chapters of this book traverse the breadth of seismic issues relevant to architects, the emphasis throughout is on designing buildings capable of resisting seismic forces both safely and economically. As designers we have a responsibility to deliver buildings to clients that, subsequent to mutually acceptable discussions about seismic performance, meet their expectations. For a code-level quake, the goal of achieving life-safety will be met and the level of structural and non-structural damage not unexpected. We definitely want to avoid an outcome where buildings that we have designed experience unanticipated serious damage, and our clients suffer heavy losses in terms of injuries to occupants and financial losses including losses due to downtime.

I have recently viewed the photographs taken by a reconnaissance team just returned from the earthquake damaged region around Pisco, Peru. Thousands of collapsed and damaged buildings, both vernacular and modern, succumbed to the ground shaking of the August 2007 quake because they lacked one or more seismic resistant features. Once buildings are damaged their seismic deficiencies in design, detailing, construction, and even maintenance, are clearly revealed to engineers and architects skilled in analysing earthquake damage.

A damaging quake like that at Pisco recalls and illustrates lessons on seismic resistant design. As we study one collapsed building we recognize an absence of seismic resistance along one of its plan orthogonal axes. Another building illustrates a lack of a dependable force path when its façade falls away from side walls to collapse in the street. Elsewhere, we observe partial or complete collapse due to soft-storey or short column configurations. With a sound understanding of seismic resistant principles we can analyse many causes of damage. But an ability to *analyse* seismic damage is just one application of seismic knowledge. Now we must apply seismic resistant principles in the process of architectural *design*.

Whereas the act of seismic analysis discussed above mainly requires focused technical knowledge, undertaking the seismic design of a building demands a far broader approach and a wider range of skills. Not only are structural systems and elements to be provided, but they necessitate thorough integration with architectural planning hopefully enhancing the building function and architectural design concepts. During the process of configuring structure, many other architectural issues demand attention. Entry, circulation, quality of interior spaces and natural light, are but a few of the issues to be addressed concurrently. And all the while coping with these architectural aspects the critical importance of achieving a sound seismic configuration must be remembered. A poor or flawed configuration will not perform as well as one based on proven seismic design principles.

So, let's assume an architect has just commenced the preliminary design of a building. Rough sketches of the building massing and one or more floor plans have been developed. Perhaps there are vague ideas of how gravity forces might be supported. So how does he or she begin a preliminary seismic design? While acknowledging that a flowchart-type design methodology is too prescriptive for a process comprising a synthesis of technical, aesthetic and spatial considerations, Table 18.1 attempts to summarize the process. The table can function as a design guide or checklist and suggests that a designer begins by attending to the primary structure and then moving onto secondary elements.

Table 18.1 should assist architects to meet the requirements for a basic code-complying design. Material from other chapters provides layers of additional seismic design sophistication and presents opportunities to refine seismic design solutions. For example, even before commencing a preliminary design, the architect and client can consider the trade-offs between ductility and strength that will affect how much vertical structure is required; and between the stiffness and flexibility of vertical structure that influences the widths of seismic separation gaps from boundaries and adjacent buildings on the same site.

Finally, a note about architect–structural engineer collaboration. Ideally, an architect should possess sufficient knowledge and confidence to undertake a preliminary seismic design alone. Then the structural system will be sympathetic to building function and other architectural aspirations. However, a professional structural engineer should be involved at the earliest opportunity. Specialist structural advice will prove invaluable in refining preliminary architectural design ideas and avoid excessive reworking of floor plans and sections if structural modifications are warranted. Where structural requirements

▼ **18.1** Summary of the seismic design process

Structural system	Questions to be asked as the preliminary seismic design is developed	Comments	Chapter
Primary vertical structure	Are there locations in plan for shear walls of sufficient length and thickness that rise from foundations to roof to resist seismic forces?	Shear walls are the best means of seismic resistance.	5
	Does *each* plan orthogonal direction have two or more shear walls, braced or moment frames?	The structural elements must be large enough to provide adequate strength and stiffness. Avoid mixed-systems and remember that stiffer systems will allow you to build closer to site boundaries and lessen widths of any seismic gaps between structure and non-structural elements.	2 and 5
	Are the structural systems off-set in plan to resist torsion?	Torsional stability must be ensured.	8
	Are there any configuration problems that need resolution?	Avoid weak columns–strong beams, soft-storeys, etc.	9
Primary horizontal structure	Are the floors and roof able to function as diaphragms in both orthogonal directions to transfer seismic forces horizontally to the vertical structure?	Flexible or highly penetrated diaphragms may require vertical structure to be more closely-spaced in plan.	4
	Are there any configuration issues such as re-entrant corners or diaphragm discontinuities?	Seismic separation of one building into several structurally independent buildings may be required.	8
Secondary structure	Do walls have enough depth and strength to transfer out-of-plane forces to diaphragms above and below?	An important consideration for heavy and high walls.	2
	Are strong non-structural elements structurally separated?	Check that structural problems will not be caused by strong infill walls or staircases.	10
	Are other non-structural elements separated from the damaging effects of interstorey drifts yet tied back to the main structure?	Prevent seismic damage to elements caused by interstorey drifts and horizontal accelerations.	11

are addressed for the first time late in the design process, structure is often poorly integrated architecturally and more expensive.

May you apply the principles of seismic design gained from this book as well as from other sources to design safe, economical and architecturally desirable buildings capable of outwitting the quake.

RESOURCES

INTRODUCTION

Although bibliographic references are provided at the end of each chapter, this section summarizes the two most significant generic sources of seismic design information for English-speaking architects. Institutions and organizations, which are often publishers themselves are listed, followed by books from established publishing houses. These lists contain the most architecturally relevant seismic resources. Below each of the listed institutions and organizations, brief comments indicate the material that is likely to be most useful to architects. Much of the material from websites can be downloaded for free. Each book reference is annotated, noting special features.

INSTITUTIONS AND ORGANIZATIONS

Applied Technology Council (http://www.atcouncil.org/)

Publications, handbooks and manuals including post-earthquake building assessment material and a continuing education publication: (1999). *Built to resist earthquakes: the path to quality seismic design and construction for architects, engineers, inspectors.*

California Seismic Safety Commission (http://www.seismic.ca.gov/)

Many reports including guidelines for owners of masonry buildings, commercial properties and homeowners.

Consortium of Universities for Research in Earthquake Engineering (CUREE) (http://www.curee.org/)

In the Earthquake Architecture section over twenty papers are downloadable.

Earthquake Engineering Research Institute (EERI)
(http://www.eeri.org/)

A monthly newsletter with reports on recent earthquakes, the journal *Earthquake Spectra*, reports, videos, CDs containing images from quake-affected regions and much more.

Earthquake Hazard Centre
(http://www.victoria.ac.nz/architecture/research/ehc/)

A quarterly newsletter written for practicing architects and structural engineers in developing countries.

Federal Emergency Management Agency (FEMA)
(http://www.fema.gov/)

Publications for US architects and engineers including procedures for rapid assessment of earthquake damaged buildings, information on school safety, techniques for seismic retrofitting and:

(2006). *Design for Earthquakes: A manual for architects* (FEMA 454). The publication aims 'to help architects and engineers become better partners, not to further their separation, and to encourage a new level of architect and engineer collaboration'. It contains in-depth and wide coverage of topics relevant to architects practicing in seismically-prone regions.

(2004). *Primer for Design Professionals: Communicating with owners and managers of new buildings on earthquake risk* (FEMA 389). Detailed guidance on managing and reducing seismic risks together with seismic design and performance issues for a range of building types and facilities.

Multidisciplinary Center for Earthquake Engineering Research
(MCEER) (http://mceer.buffalo.edu/)

Comprehensive earthquake information for design professionals. The QUAKELINE® database is essential for those engaged in earthquake-related research and EQNET is a web portal to other earthquake resources.

National Geophysical Data Center (NGDC) (http://www.ngdc.
noaa.gov/)

Over thirty slide sets illustrating earthquake damage from around the world.

National Information Center of Earthquake Engineering- India (NICEE) (http://www.nicee.org/)

Educational resources and many publications including architectural teaching resource material on seismic design concepts for teachers in Indian architecture colleges, and:

(1986). *Guidelines for earthquake resistant non-engineered construction.* International Association for Earthquake Engineering (IAEE).

Bachmann, H. (2003). *Seismic conceptual design of buildings—basic principles for engineers, architects, building owners, and authorities.*

National Information Service for Earthquake Engineering (NISEE) (http://nisee.berkeley.edu/)

As well as a library dedicated to earthquake engineering, the Center provides numerous earthquake engineering information resources available over the internet including an open access archive, slide sets, images and photographs, selected full-text papers and related websites.

New Zealand Society for Earthquake Engineering (http://www.nzsee.org.nz/)

Massey, W. and Charleson, A. W. (2007). *Architectural design for earthquake: a guide to the design of non-structural elements.*

World Housing Encyclopedia (http://www.world-housing.net/)

The encyclopaedia is a resource for sharing knowledge on construction practices and retrofit techniques. With an emphasis on developing countries, aspects of housing construction from over thirty countries and 100 housing types are included, such as socio-economic issues, architectural features, structural systems, seismic deficiencies and earthquake resistant features, performance in past earthquakes, available strengthening technologies, building materials and the construction process.

Several 'tutorials' are directly relevant to practicing architects and engineers, such as 'At Risk: The seismic performance of reinforced concrete frame buildings with masonry infill walls', and others on masonry and adobe house construction.

US Geological Survey (USGS) (http://www.usgs.gov/)

Earthquake fact sheets and summary posters, shaking hazard and intensity maps for the US including liquefaction hazard information.

PUBLICATIONS

Ambrose, J. and Vergun, D. (1999). *Design for Earthquakes*. John Wiley & Sons. Partially detailed design cases to the Uniform Building Code (UBC) for the following building types: wood-framed residence, single-storey commercial, low-rise multi-unit, low-rise office, multi-storey apartment and single-storey warehouse.

Arnold, C. and Reitherman, R. (1982). *Building Configuration and Seismic Design*. John Wiley & Sons. The classic text for architects on building and structural configuration from the perspective of seismic design. Still relevant twenty-five years later.

Booth, E. and Key, D. (2006). *Earthquake Design Practice for Buildings* (2nd Edn). Thomas Telford Publishing. Written for practicing structural engineers, the practical approach of the book and its numerous figures makes it accessible to architects. Addresses US and Eurocode requirements.

Coburn, A. and Spence, R. (1992). *Earthquake Protection*. John Wiley & Sons. A non-technical overview of earthquake risk reduction strategies for societies including a chapter on improving the seismic resistance of buildings.

Dowrick, D.J. (1987). *Earthquake Resistant Design for Engineers and Architects* (2nd Ed.). John Wiley & Sons. Although containing earthquake engineering detail for engineers some construction details and chapters on structural form and architectural detailing are of relevance to architects.

Kuroiwa, J. (2004). *Disaster reduction: living in harmony with nature*. Editorial NSG S.A.C., Peru. Responds to Peru's and other developing countries' need to reduce the effects of disasters including earthquakes. Contains extensive information on the materials, construction and seismic resistance of buildings common in developing countries. Reports on lessons learned from laboratory tests and earthquake damage.

Lagorio, H.J. (1990). *Earthquakes: An architect's guide to non-structural seismic hazards*. John Wiley & Sons. A broad consideration of seismic hazards relevant to US conditions. Chapters on earthquake recovery and reconstruction and earthquake hazards mitigation processes.

Naeim, F. (ed.) (2001). *The Seismic Design Handbook* (2nd Edn). Kluwer Academic Publishers. Detailed treatment of earthquake engineering, primarily for engineers. Written for U.S. codes and design practice.

Chapters on 'Architectural considerations' by Christopher Arnold and another on the design of non-structural systems and components are of relevance to architects.

Schodek, D.L. (2005). *Structures* (5th edn). Pearson Prentice Hall. A general text on structures for architects in which the dynamic characteristics of seismic forces are briefly introduced and earthquake design considerations are included in a chapter on designing for lateral forces.

Stratta, J.L. (1987). *Manual of Seismic Design*. Prentice-Hall. Selection of images show earthquake damage to buildings from thirteen different quakes. Further images illustrate damage to architectural elements and building contents.

INDEX

Accelerations:
    amplification by superstructure, 11,
      12, 13, 14, 22, 159, 166, 184
    horizontal, 11, 12
    vertical, 11, 18, 218
Accelerogram, 21
Adjacent buildings see Pounding
Aftershocks, 8
Algeria, 38, 245
Alpine fault (New Zealand), 8
Ando, Tadao, 252
Appendages, 181
Applied Technology Council, 269
Architectural conservation, 203
    plan, 204
Architectural design concept, 56, 94,
    96, 208, 211, 259
Architectural integrity, 188
Architecture:
    contemporary, 104–8
    integration with structure, 93–9
    Arnold and Reitherman, 37, 148
Arnold, Christopher, 125, 211, 252,
    258

Bahrainry, Hossein, 236
Bandaging, 256
Base isolation see Seismic isolation
Basements, 122
Bendimerad, Fouad, 245
Bending moment, 24, 32, 51, 257
Bhuj earthquake, 243
Binding, 256
Bond beams, 30–1, 58–60
Botta, Mario, 105
Boumerdes, Algeria earthquake, 245
Boundary walls, 42
Braced frames, 66–7, 76–81, 130,
    200–1, 257
    ductility, 80

eccentrically braced, 78
materials, 79
structural requirements, 78–9
tension and compression, 77, 80
tension-only bracing, 77
types of, 76–8
Bridges between buildings, 140
British Columbia, 190
Brittleness, 23–4
Bryant, Edward, 237
Buckling-restrained brace, 226
Building:
    design life, 9, 39
    flexibility, 13
    height, 100
    performance, 208
    stiffness, 26
    strength in two directions, 26
    weight, 100
Building configuration see Horizontal
    configuration and Vertical
    configuration
Building contents, 3, 40, 184–6, 220,
    230, 255
Buried structures see Underground
    structures
Business continuance, 158, 185, 209,
    217, 221, 229–30

California, 8, 34–7, 187, 190, 234, 236,
    238, 251
California Seismic Safety Commission,
    269
Capacity design, 35–6, 43–6, 53, 58,
    145, 153, 189, 217, 247, 256–7
    foundations, 119
    moment frames, 82, 86–9
    shear walls, 73–6
Captive column, 150
Carbon fibres, 230

Cast-in-place concrete:
  topping slab, 53
CCTV Headquarters, 107, 231, 262
Centre of mass, 16, 27, 128–32,
  153–4, 162
Centre of rigidity or resistance, 27–8,
  128–32, 152–4, 162, 169
Chicago School, 37
Chile, 9
China, 9, 107, 231
Chords, 52, 54
Circulation, 95
Civil defence, 3, 40
Cladding:
  brick veneer, 175–6
  masonry, 174–6
  panels, 106, 152, 177–9
Client, 207, 208–10, 230, 239, 246–7
Coburn and Spence, 234
Codes of practice, 3, 8, 23, 25–6,
  34–6, 38–9, 126, 131, 133, 137,
  144, 163, 173, 175, 185, 189, 191,
  208, 221, 240
  design-level earthquake, 39–42, 245
  minimum standards, 39
  objectives, 39–40
Collaboration, 207–16, 266–7
Collectors and ties, 61, 132, 258
Column stiffness, 148
Comerio, Mary, 230
Communication, 207–16
Configuration see Horizontal
  configuration and Vertical
  configuration
Confined masonry walls, 69–73, 163,
  246
Conservation, 203
Consortium of Universities for
  Research in Earthquake
  Engineering, 269
Construction practice, 213
  quality, 213, 248
Continental drift, 1
  see also Tectonic plates
Contractors, 297, 231–5
Cook, Peter, 98
Corner building, 131–2

Costs:
  earthquake damage, 158, 209, 217,
    229
  non-structural damage, 158
Cost of earthquake resistance,
  3, 210
  factors influencing cost, 3, 42
Coupled shear walls see Shear walls
Critical facilities, 40
Curtain-walls, 179–81

Damage avoidance systems, 227–8
Dampers, 21, 34, 255
  active, 226–7
  passive, 224–6
Damping, 20–1, 24, 219
Design concept see Architectural
  design concept
Design-level earthquake, 8, 39–42,
  209, 245
Design team, 210–12
Developing countries, 3, 168, 243–9
Diagonal compression strut,
  150, 161
Diaphragms, 50–8, 193, 258, 267
  ceiling, 30
  chords, 52–3
  connection details, 51–2
  continuous, 53
  cross-braced, 55
  discontinuities, 127, 134–6
  flexible, 52
  kinked, 135–6
  materiality, 53–4
  penetrations, 56, 99, 133–4
  retrofit, 198
  rigid, 52
  rotation, 129–30
  transfer, 56–8, 152, 154
  trusses, 54
Disaster management, 3
Discontinuous walls, 151–3
Double-height floor, 147
Drift see also Interstorey deflection
Ductility, 23–5, 35, 41–6, 130, 147,
  189, 207, 214, 266
Duration of shaking, 10, 12, 13–14

Early warning systems, 2
  tsunami, 238
Earthquake *see also* Seismic
Earthquake architecture, 104,
  251–68
Earthquake directivity, 11, 13
  randomness, 13, 65
Earthquake ground shaking:
  characteristics of, 4, 12–14
  influence of soft soil, 13–14
Earthquake Hazard Centre, 248, 270
Earthquake Engineering Research
  Institute, 270
Earthquake intensity, 9–10, 210
Earthquake magnitude, 6, 9
Earthquake occurrence:
  explanation, 2, 4
  frequency, 7–8
  locations, 7
  probability, 9, 209
Earthquake prediction, 7–8
Earthquake proof, 207, 224, 228
Earthquake resistance:
  cost of, 3
  how to achieve it, 3
Earthquake return period, 8–9, 39–40,
  245
Earthquake scenario, 210
Earthquake shaking:
  directionally random, 26, 31
  experience of, 2–3
  frequency content, 12, 13
  torsion, 128
Earthquake waves, 11–12, 19, 258,
  260
  frequency content, 12, 13
  Love waves, 11
  primary (p-) waves, 11
  Rayleigh waves, 11
  shear (s-) waves, 11
  surface waves, 11
Earth's crust, 4
Educative role, 208
Eisenman, Peter, 258
Elastic response, 42
Electrical equipment, 184
Elsesser, Eric, 90, 228

Epicentral distance, 9, 10
Epicentre, 6, 9, 10

Face-loads *see* Out-of-plane forces
Fardis, Michael, 163
Farquharson, David, 251
Fault or fault-line, 6, 259–60
  length, 6
  maps, 234
  surface rupture, 118, 236–7, 253
Federal Emergency Management
  Agency, 270
Fernández-Galiano, Luis, 262
Fintel, Mark, 66
Fire following earthquake, 238–40
Fire rating, 182
Fire stations, 8, 40, 191, 234
Flashings, 106, 138, 141
Flat plate/slab, 84
Floor penetrations *see* Diaphragms
Flotation, 122
Focal depth, 6
Focus, 6
Force paths, 29–31, 49, 107, 144, 193,
  265
Folding, 260
Foreshocks, 8
Foundations, 25, 113–23, 195
  failure, 45, 118
  investigations, 119–21
  problems and solutions, 114–19
  shear walls, 68
  types, 119
Fracturing, 260
Fragmentation, 259
Free-plan, 37
Frequency content of ground shaking,
  12, 13
Friction pendulum dampers, 221
Fuse *see* Structural fuse

Geotechnical engineer, 113–14
Glazing, 180
Gravity resisting structure, 96–8, 146,
  194, 254, 260
Ground anchors, 68, 121
Ground improvement, 116–17

Guevara and Garcia, 150
Guevara-Perez, Teresa, 240
Gülkan, Polat, 243

Hawaii, 238
Hermès building, 33, 119
Historic buildings, 188, 223
History of seismic design, 33–8
Horizontal configuration, 125–41, 267
  irregularities, 127–37, 191, 193, 246
  regularity, 126
Horizontal structure, 49–58, 98–9,
  267
  bond beams, 58–60
  collectors and ties, 61
  see also Diaphragms
Hospitals, 8, 40, 113, 158, 184, 191,
  234–5
  Olive view, 37, 220
  Southern California Teaching, 219

Improvement see Retrofitting
India, 239, 243–4, 248
Inertia forces:
  building weight, 18
  damping, 20–1
  ductility, 23–25
  factors affecting their severity,
    18–25
  nature of, 15–16
  natural period of vibration, 18–23
Infill walls: 159–68
  masonry, 37, 69–72, 146, 165, 174,
    202, 246
  partial-height, 148, 151, 167
  problems, 160–2
  retrofit, 202–3
  solutions, 162–8
Informal buildings see Non-
  engineered
Insurance, 190, 192, 210, 221
Interstorey deflection, 40, 66, 90, 106,
  137, 147, 159, 162, 165, 169,
  177–80, 194, 224, 255, 267
Intraplate earthquakes, 7
Inundation map, 237
Iran, 6, 236

Isoseismal map, 10
Isozaki, Arata, 262
Istanbul, 8, 236
Italy, 34

Jain, Sudhir, 243
Japan, 34, 36, 38, 104–5, 163, 218, 220,
  261
  Kashiwazaki nuclear plant, 237
  Sanriku coast, 237

Kobe earthquake, 117, 122, 128,
  145, 190, 218, 219, 239–40, 252,
  261–2
Kuroiwa, Julio, 249

Laboratories, 185, 230
Landslides, 113, 118
Langenbach, Randolph, 204
Lateral spreading, 113, 117
Latin America, 249
Lead-extrusion damper, 225, 255
Lead-rubber bearings, 221
Le Corbusier, 37–8, 108
Lifelines, 234–6
Liquefaction, 114–17, 234
Load-bearing wall, 63
Load paths see Force paths
Loma Prieta earthquake, 89, 190
Look, Wong and Augustus, 188, 203
Los Angeles, 104, 158, 187, 189

Masonry construction, 58
Masonry infill walls see Infill walls
Mayes, Ron, 221
Maximum credible earthquake, 8
Mechanical systems, 97, 99, 210, 223
  restraint, 184
Metaphors, 259–60
Mexico City, 2, 13
Microzone maps see Seismic maps
Mixed systems, 89–91
Modes of vibration, 18–20
Modified Mercalli Intensity Scale,
  10–1, 210
Moment frames, 64–6, 81–90, 95, 130,
  201, 254, 256

ductility, 83, 86–9
essential characteristics, 81
materials and heights, 85–6
mixed system, 90
structural requirements, 83–5
Moment of inertia, 27
Moss, Eric Owen, 104
Multidisciplinary Center for
    Earthquake Engineering
    Research, 270
Museum of New Zealand, 258

National Geophysics Data Center,
    270
National Information Center of
    Earthquake Engineering (NICEE),
    248, 271
National Information Service for
    Earthquake Engineering, 271
Natural period of vibration, 18–20,
    22–3, 26, 160, 219, 253
Newcastle, Australia, 175
New technologies, 217–31
Newton's second law of motion, 16,
    34
New Zealand, 38, 163, 175, 187,
    190–1, 218, 222, 223, 236, 256
Nishitani, Akira, 227
Non-ductile, 24
Non-engineered buildings, 244–5
Non-parallel systems, 127, 136–7
Non-structural elements, 20, 26, 40,
    42, 152, 209, 211, 254–5, 267
    causing structural damage, 157–70
    not causing structural damage,
        173–86
    restraints, 159, 184
    retrofit, 202–3
Northridge earthquake, 6, 8, 36, 51,
    90, 117, 128, 158, 179, 188, 190,
    219, 239, 261
Nunotani Headquarters building, 258

Off-set walls, 151–3
OMA, 107
Out-of-plane forces, 30–1, 51, 162,
    164–5, 177–9, 181, 193, 195, 267

Pakistan, 2
Parapets, 181
Partition walls, 182
    fire-rated, 182
Peak ground acceleration, 12–13, 22
Performance-based design, 229–30
Peru, 236, 265
Philosophy of seismic design, 34–42
Piles, 115, 117
    tension, 68, 79
Planning see Urban planning
Plasterboard, 52–3
Plastic hinge, 24, 41–6, 68, 88, 99, 150,
    254, 256
Plywood, 52–4
Poisson's equation, 9, 39
Post-and-beam, 63
Post-earthquake:
    assessment, 215
    emergency response, 235
    fires, 235
    reconnaissance, 8
    redevelopment, 235
Pounding, 137–9, 195, 253
Practical columns, 165
Precast concrete:
    damage, 243
    damage avoidance, 227
    flooring, 53
Preliminary architectural design
    models, 96

Quality of construction, 213–4, 248

Raised floors, 182–3
Redundancy, 87, 90–1, 97, 130
Re-entrant corner, 127, 132–3
Rehabilitation see Retrofitting
Reinforced concrete, 12
    over-reinforced, 44
    shear failure, 44, 46
Renzo Piano Building Workshop, 33,
    119
Rescue activities, 8
RESIST, 102
Resonance, 20–2
Response spectrum, 21–3

Restraints, 184
Retaining structures, 121–2
    basements, 122
Retrofitting, 18, 37, 187–205, 235, 252
    adjacent buildings, 195
    approaches, 192–5
    assessment, 190–1
    historic buildings, 203–4
    non-structural elements, 202–3
    objectives, 191–2
    reasons for, 189–90
    techniques, 195–203
Return period, 8
Richter scale, 9
Rocking, 119
Roof band see Bond beam
Russia, 238

San Fernando earthquake,
    179, 239
San Francisco, 8, 34, 138, 188–9, 234
    International Airport, 221
    Museum of Modern Art, 105–6
Saunders, Mark, 138
Scawthorn, Charles, 239
Schools, 40, 243
Seattle Public Library, 107
Seismic see Earthquake
Seismic design:
    force, 23
    history of, 33–8
    philosophy of, 34–42
Seismic forces see Inertia forces
Seismic isolation, 218–24, 255
    retrofitting, 202
Seismic joints see Separation gaps
Seismic map, 9
    hazards, 234–5
    vulnerability, 235
Seismic zone, 100
Seismograph, 9, 10, 12, 260
Separation gaps, 128, 133, 137–9,
    146–7, 155, 159, 163, 177–80,
    195, 203, 211, 214, 222, 252–5,
    267
Separation of gravity and seismic
    systems, 97–8, 146

Services (see also Mechanical
    systems), 97, 115
Setbacks, 154–5
Shape memory alloys, 230
Shear failure:
    reinforced concrete, 44, 46
Shear forces, 25, 32, 51, 257
Shear walls, 64–76, 95, 130, 163,
    198–200, 246, 256–8, 267
    coupled, 75–6
    ductility, 73–5
    materials, 66, 69–73
    mixed system, 90, 267
    penetrations, 68
    sloped, 67
    structural requirements, 68–73
Shell structures, 102–4
Shock absorbers see Dampers
Short columns, 148–51, 191,
    246, 265
Shotcrete, 198
Sinha and Adarsh, 244
Slicing, 259
Sliding, 260
Sliding joints, 140, 169–70, 254
Slope instability, 234
Sloping site, 149, 155
Soft storey, 91, 144–9, 161, 163, 191,
    246, 254, 265
Soil:
    failure, 46
    influence of, 10, 13, 23, 101
    natural frequency, 15
    shaking amplification, 13, 114, 220
    soft, 10, 14, 93, 234
    Spandrel beams, 147–8
Spandrel panels, 147
Splitting, 260
Staggered wall, 153
Staircases, 168–70, 203
Strapping, 256
Strengthening see Retrofitting
Structural design approaches, 229–31
Structural footprint, 42, 93, 101, 207
Structural fuse, see also Plastic hinge,
    24–5, 38, 41–6, 53, 78, 86–7, 99,
    153, 214, 255, 257

Structural principles and actions:
    expression of, 255–8
Subduction, 4
Subsidence, 117
Sumatra earthquake, 237
Surface fault rupture, 118
Suspended ceilings, 3, 182

Taiwan, 20
Teamwork, 208
Tectonic plates, 1, 4–5
    continental plate, 4
    location, 5
    movement, 4–7
    oceanic plate, 4
    Tension membranes, 102
Timing of structural design, 96, 208,
    211, 266
Tokyo, 2, 33, 109, 239
    Imperial Hotel, 228, 253
Torsion, 11, 16, 27–8, 65, 97, 127,
    128–32, 152, 162, 169, 193, 246,
    267
Torsionally unbalanced system, 131
Transfer diaphragms, 56–8, 152, 154
Trusses:
    horizontal, 54
    vierendeel, 55
Tsunami, 9, 237–8
Turkey, 153, 240, 243

Underground structures, 122
Uniform Building Code, 35
Unreinforced masonry:
    buildings, 51, 187
    parapets, 181
    walls, 195–8
Upgrading see Retrofitting

Urban planning, 233–40, 253
USA, 38, 163, 180, 191, 193, 213, 218,
    223, 227, 228, 261
US Geological Survey, 271
Uzbekistan, 244

Veneer, 175–6
Venezuela, 127, 246
Vernacular architecture, 33
Vertical accelerations:
    need to design for, 18, 218
Vertical configuration:
    discontinuous walls, 151–3
    irregularities, 143–55, 191, 193,
        246, 267
    off-set walls, 151–3
    short columns, 148–51
    sloping site, 149
    soft storeys, 144–8
Vertical structure, 63–91, 98–9, 193,
    267
    how much is needed, 99–102
    number of elements, 101, 130
Vibroflotation, 116
Vierendeel frame, 134–5
Villa Savoye, 108–11

Waffle slabs, 84
Weight reduction, 18, 194
Wellington, 189, 192,
    Victoria University library, 223–4
Whiplash accelerations, 20
Wind force, 17, 51, 93
Windows, 179–81
Woods, Lebbeus, 261
World Housing Encyclopaedia, 248,
    271
Wright, Frank Lloyd, 228, 253

Milton Keynes UK
Ingram Content Group UK Ltd.
UKHW051926141024
449569UK00027B/1376